Dedication

This book is dedicated to the ones who mean so much to us. The ones who make us laugh, make us cry, the ones we love.

To all our children, Patrick and Donna, Kimberly and Tom, Robert and Staretta, Dawn, Darci, Amber and Chet, Michael, and especially to Harley Wilkins, and to our grandchildren, Megan Murphy, Danielle Azevedo, Matthew McNairy, Samantha McNairy, Emily Murphy, and Kyle Azevedo, for their part in this book.

Thank you for overlooking our shortcomings and for loving us and understanding when we didn't have time.

This book is for you.

HOME VCR REPAIR ILLUSTRATED

Richard C. Wilkins
Vicki Wilkins

Second Edition

McGraw-Hill

New York   San Francisco   Washington, D.C.   Auckland   Bogotá
Caracas   Lisbon   London   Madrid   Mexico City   Milan
Montreal   New   Delhi   San Juan   Singapore
Sydney   Tokyo   Toronto

Library of Congress Cataloging-in-Publication Data

Wilkins, Richard C.
    Home VCR repair illustrated / Richard C. Wilkins, Vicki Wilkins.—
2nd ed.
        p.   cm.
    ISBN 0–07–070769–3
    1. Videocassette recorders—Maintenance and repair—Amateurs'
manuals. I. Wilkins, Vicki II. Title.
TK9961.W56   1999
621.388′337—dc21   99–35899
                                                                      CIP

McGraw-Hill

*A Division of The **McGraw·Hill** Companies*

1 2 3 4 5 6 7 8 9 0 AGM/AGM 9 0 9 8 7 6 5 4 3 2 1 0 9

ISBN 0-07-070769-3

*The sponsoring editor for this book was Scott Grillo and the production su-
pervisor was Pamela Pelton. This book was set in Melior by Jan Fisher
through the services of Barry E. Brown (Broker—Editing, Design and
Production).*

Printed and bound by Quebecor Martinsburg.

McGraw-Hill books are available at special quantity discounts to use as
premiums and sales promotions, or for use in corporate training programs.
For more information, please write to the Director of Special Sales,
McGraw-Hill, 11 West 19th Street, New York, NY 10011. Or contact your
local bookstore.

CONTENTS

PREFACE

I WAS SITTING IN MY SHOP FIXING A VCR WHEN MY FIANCEE BEGAN opening the day's mail. She stopped on one particular letter and read it out loud to me. She was so excited I had to tell her to slow down. It was a letter from McGraw-Hill wanting to publish a second edition of my book, "*Home VCR Repair Illustrated*". We discussed the pros and cons of this endeavor and decided that, yes, people could be helped by an updated version of this book.

I had changed locations and situations, being newly engaged (at my age, that's an extreme change). I decided to call McGraw-Hill and let them know of our decision. I say "our" because my fiancee has taught this "old dog" that in a working relationship, *everything* has to be discussed. A few weeks later I received and signed the contract to begin the new "*Home VCR Repair Illustrated* – Second Edition". And so began the new adventure with my fiancee and soon to be wife (don't let anyone tell you that it's easy to work with the one you love 24 hours a day). This book is the finished product of a "labor of love".

The goal I have for this book is to teach anyone how to avoid spending $100 for a $10 job. At a supposedly reputable shop, to blow the dust out, install a maintenance kit, clean the video heads, clean and lubricate the unit, it ran the consumer about $110. With the aid of this book you can do the same job for less than $15. This book will save you money on the cost of parts and eliminate the cost of all labor.

The icing on this book is that I've included all my trade secrets, eliminated all testing equipment, showing you how to use common household items to assist you. All technical aspects of electronics are presented through my experiences in repairing VCRs since they were first introduced to the consumer. When using a manual, test equipment, and technical theories, a roller guide alignment takes about two hours on the average. Using my method takes about 15 minutes, without any testing equipment. There's no difference in accuracy from their method to mine, just a lot of time and money.

ACKNOWLEDGMENTS

Thank you to Scott Grillo, Acquisitions Editor, McGraw-Hill, for his patience and understanding in giving me the time to complete this manuscript.

Thank you to Kelly Ricci and her team for the excellent job in editing the book.

Thank you to Loretta Yates, Managing Editor for Howard W. Sams & Company, for her approval to reference this "must have" annual index.

Thank you to John Dwinell, Sr., President of PRB Line Electronic Products, for approving the use of this very helpful cross-reference catalog.

A very special thank you to Max McCarty, owner of Pacific Photo Lab in Eugene, Oregon, for his expertise and excellent quality of work.

To J.H. and Francis Wilkins (mom & dad), thank you for your moral support and giving ways. You were always there.

To Mike and Reita Brown, thank you for the use of your video equipment and the friendship you have shown me over the years.

A very, very, special thank-you to my co-author and beautiful wife, Vicki, without whose support and love, I never could have finished this project. Thank you for the late hours, the early mornings, and most of all the encouragement and faith you had in me.

INTRODUCTION

This book was written based upon my ideas and experience in VCR repair.

The method is simple and safe to use on any VCR, if you follow the directions carefully. It's designed to teach you how to do-it-yourself in your own home, using household items and basic tools. No test equipment is involved. This book gives you step-by-step directions.

You will probably find it helpful to refer to Chapter 20 and diagnose the problems in your VCR. This book is broken down into sections that correspond to each particular section of a VCR so you can proceed to the appropriate section.

All chapters contain a review section. This section provides step-by-step instructions to assist in diagnosing and repairing the VCR. This section also acts as a checklist so you don't miss any steps when performing repairs.

Each chapter contains referrals for further information in other chapters, to give you the best understanding possible. This book zero's in on a problem, locates it, and shows how to remove, repair, and replace any part. Besides providing information on repair, this book contains knowledge not previously accessible to anyone.

Tools and Supplies Needed

If you use a VCR as much as I do, you know how frustrating it can be to watch a distorted picture or be stuck without a picture. To top it off, you enjoy and take pride in fixing things yourself and would fix your VCR, if only you knew how. All it takes to save a lot of time and money is some basic information. After you realize how simple and inexpensive it is to do repairs yourself, you'll probably wish this book were available a long time ago.

Eventually, a VCR will require cleaning and lubricating—especially if you rent a lot of videocassettes. There's no reason to pay a service charge when routine cleaning and maintenance are required. Just a little preventative maintenance can save you money and prevent costly repairs in the future.

Tools

The tools you will need to begin are listed here. These items can be purchased at your local hardware store.

- a small-tip and medium-tip magnetic Phillips screwdriver, (if you don't already have a magnetic Phillips screwdriver, don't purchase one, I'll show you how to magnetize a Phillips head)
- two flathead screwdrivers, one standard medium size tip and one with a very small tip
- a 1.5-mm, a 1.27-mm, or 0.89-mm Allen wrench
- a 5⁄16-inch socket, 7⁄32-inch socket, and 1⁄4-inch socket, or the same size of nut drivers
- long-nose pliers
- wire cutters
- a soft-bristled paintbrush
- super glue
- a clear silicone rubber sealer

Materials

Because you're just starting out, here's an important shopping list of essential supplies:

- one package of chamois sticks
- head cleaner or cleaning alcohol
- a tube of phono lube
- a container of tuner grease
- a spray can of degreaser (the degreaser must be the type that completely evaporates. Check for complete evaporation by spraying it on your fingers and rubbing them together. The spray should leave no film or residue).
- a VHS Vu-Thru blank cartridge, part number PRB# V-511 (shown in Figure 1.1)
- 24-gauge stranded wire
- solder wick
- rosin-type solder
- a soldering iron or gun
- a C-ring puller
- an alignment screwdriver, (see Figure 1.3)
- black electrical tape and an oilier lubricator with a precision tip that fits into hard-to-reach places.

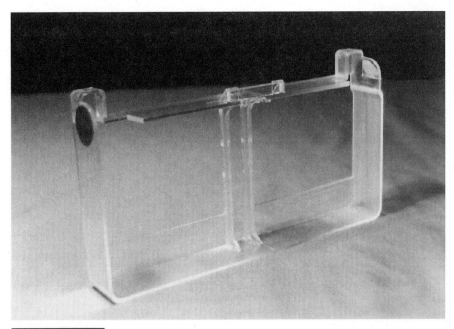

FIGURE 1.1

Blank cartridge.

All of these items can be purchased at your local electronics part store, (see Figure 1.2 for examples) or you can refer to the "Parts" section in this chapter.

You should also have on hand:

■ a roll of masking tape, ¾-inch clear scotch tape

■ a single-edge razor blade

■ an old toothbrush

■ a sharp pair of scissors

■ paper towels

■ a can of non-spraying household oil

■ two nut-digging tools or two cuticle-remover nail-cleaning tools

■ an ice pick and a clean dry empty lid or container.

These items can be purchased at your local drug store. Last, but not least, you'll need a TV. The television will act as a monitor, helping you diagnose problem areas in your VCR.

FIGURE 1.2

Chemicals and supplies needed.

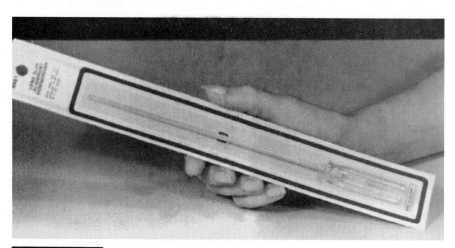

FIGURE 1.3

An alignment screwdriver.

Parts

To order parts for your VCR, pick up a Sams service manual for your machine. This book will give you all the information on service data for VCRs including the location and phone number of all authorized part distributors for each state including Alaska, Hawaii, and Canada. This book also gives you information for replacement parts and accessories for all makes and models of VCRs from coast to coast. Each company has a toll free number listed in the book. To receive this book, call 1-800-428-7267, or order on-line at, **http://www.hwsams.com.**

Another book, *PRB Line*, is a cross-reference guide and part finder. This book will show you a picture of each part and give you a description of the part and the part number. Also, you can look up part numbers by using the make and model number of your VCR. You will find 90% of the parts needed for repairs listed in this book, including alignment tools, belts and wheels, chemicals, clutches, DC motors, E-rings, fuses, gears, idler arm assemblies, idler wheels, LEDs, O-rings, pinch rollers, soldering accessories, tape sensors, tension springs, tires, tools, video heads, video drums, and more. You can obtain this book free at your closest authorized parts distributor, call 1-800-558-9752, order online at **http://www.prbline.com**, or write to info@prbline.com.

Other Items

A few other items are handy when performing VCR repairs. Because these items aren't found in the typical toolbox, I've provided a few "Trade Secrets" to save you a few bucks.

QUICK»TIP

When ordering the part, give them the make and model number of the unit, then the description and the location of the part needed. If there are any part numbers on the part give this to them. Have a credit card ready.

TRADE SECRET

MAKING A GLASS BRUSH

A glass brush simplifies repairs and is a useful cleaning utensil. Unfortunately, a glass brush isn't ready made. To make one, you'll need a fiberglass shaft or an alignment screwdriver (part numbers GC-8988, GC-8987 or GC-8728), shown in Figure 1.3. Follow these steps to transform it into a fiberglass wand:

1. First cut off the tip making it flat and flush.

2. Next heat an electric stove or hot plate to medium heat. Don't let the element get red hot. Place the cut off end onto the element.

3. Push down the wand while twisting it back and forth until a white brush appears, as shown in Figure 1.4. If an electric stove or hot plate isn't available, you can use a hot solder iron to produce the white brush, (see Figure 1.5).

You've just made a pure glass brush!

FIGURE 1.4

Making a glass brush.

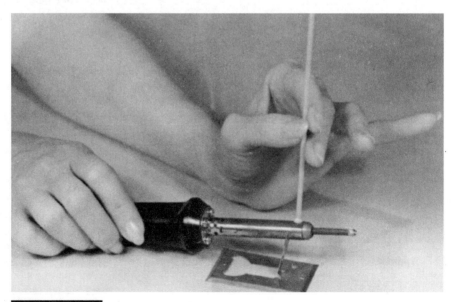

FIGURE 1.5

Using a soldering iron to make a glass brush.

MAKING A SLOTTED SCREWDRIVER

A tape-guide post tool (part number RCA# 144389 or PRB# AT389) and an audio and control head tool (part number RCA# 148936 or PRB# AT936) are used for most roller guide post and FM adjustments. Any pro will tell you that a homemade slotted screwdriver will work just as well. To make a slotted screwdriver, take a thin blade flathead screwdriver and file a notch in the center of the blade, as shown in Figure 1.6.

MAGNETIZING A SCREWDRIVER

When magnetized, a screwdriver is the ideal tool to replace and retrieve those hard-to-get-at screws. To turn an ordinary screwdriver into a magnetized screwdriver, simply rub the tip of the screwdriver across a magnet (any magnet will do). The screwdriver now is magnetized for a short period of time.

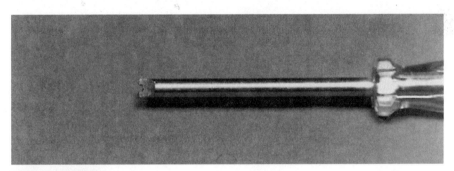

FIGURE 1.6

Screwdriver made into an adjustment tool.

Information On Screws

When removing the screws, notice the type of thread on each screw. If you place a screw in the wrong hole, the threads can be stripped. Fine-threaded screws go into a metal casing. Coarse screws go into a plastic casing. This is the general rule. The exception to this rule is the top shield plate cover over the tape path.

QUICK»TIP

If you accidentally drop a screw inside the unit, retrieve it by turning the unit upside down and shaking it, or take the unit apart. To avoid this inconvenience, use a magnetic screwdriver. When replacing a screw, the magnetic field holds it securely to the screwdriver.

Conclusion

The next chapter deals with electrical shock. Read it carefully!

Getting Inside a VCR

Before any repairs can be done on your VCR, you must first open it up to get to the source of the problem. For the first-time VCR handyman, this part of the repair process can be the most intimidating. Each time you repair your VCR, you will gain confidence; soon you'll be disassembling your VCR like an experienced repairman.

Five Ways to Remove the Main Top Cover

The main top cover must be removed to get inside a VCR. There are three different ways a top cover might be mounted. In the majority of all models, a Phillips or flathead screwdriver is required. After removing the top cover, sometimes the unit will act strange because of the light in the room; refer to the section in this chapter, "Problems that can occur after removing the top cover". When remounting the top cover, refer to the section in Chapter 1, "Information on screws."

FIGURE 2.1

First location of cover screws.

1. In fifty percent of front-load models (Figure 2.1), the top cover screws are on both sides of the unit. These four screws fasten the top cover to the chassis. Remove these top cover screws and grab hold of the cover on the bottom of each side, lifting up, then pulling straight back. This procedure releases the front lip, as shown in Figure 2.2. Now the cover lifts straight up and off. In these models, the top and both sides come off in one piece. To remount the top cover, insert the front lip under the face plate, lower the top cover, and replace the screws.

2. In some models, two screws are mounted on the back lip of the top cover on each side, as shown in Figure 2.3. After removing these screws, slide the cover straight back to release the latches. Usually two to four hidden L-shaped latches are on the inside of the main cover, holding the top cover down. Slide the top cover back about ½ inch and pick up on the sides, pulling it straight up and off. If the top cover doesn't slide back, proceed to method #3.

3. A few units only have one screw on the back lip, located in the middle of the lip on top. Place the unit on its side and look at the screws on the bottom cover plate. See if there are screws with arrows pointed at them, as shown in Figure 2.4. One screw could be on

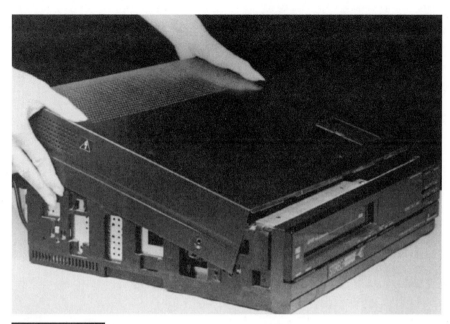

FIGURE 2.2

Releasing the front lip.

FIGURE 2.3

Second location of cover screws.

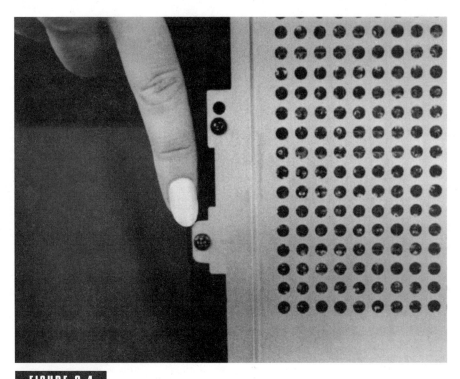

FIGURE 2.4

Arrow pointing at the cover screw.

each side, one on one side, and two on the other, or two on each side. Remove only the screws with arrows pointed at them. Then place the unit back right side up. Lift up on the sides of the top cover and slide the cover back about ½ inch and lift the cover off.

After removing the top cover, you can see the hidden latches on the chassis, as shown in Figure 2.5. When remounting the top cover, align the latches on the chassis with the slots on the bottom lip of the top cover, as shown in Figure 2.6. Place the slots over the latches and slide the cover forward to secure the latches and re-place all the screws.

4. In a very few models, you might find two screws on the back lip of the top cover on each side and a latch in the center of the back near the top, as in Figure 2.7. Remove the screws, push in on the latch, and lift the top cover off.

When remounting the top cover, be sure to align any slots or protruding latches on the bottom lip of the top cover to the match-

FIGURE 2.5

Hidden top cover latch.

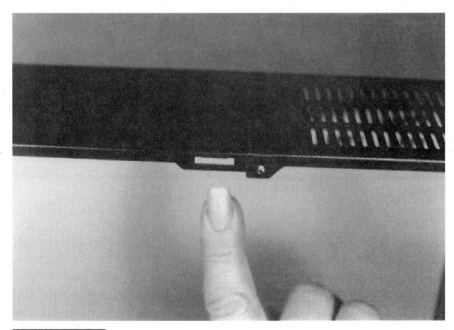

FIGURE 2.6

Slot on bottom lip.

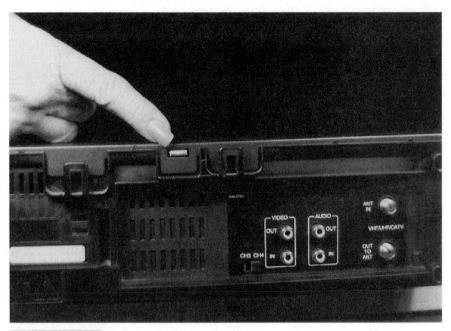

Latch in center of back.

ing slots or latches on the chassis. These slots and latches will hold the sides of the top cover in. Also, do not forget to insert the front lip under the face plate.

5. In a very few models, no screws are on the sides or the back of the unit. In this case, turn the unit over and remove any screws along the edges of the bottom plate. This will loosen the bottom plate and remove the top cover with the same screws. Turn the unit right side up and remove the cover. The bottom cover will stay attached to the unit until you remove the clips, which hold it on or remove other screws not on the edges of the bottom plate.

Problems That Can Occur
After Removing the Top Cover

After removing the main top cover, the outside room light will often penetrate one of the tape sensors and stop the unit from going into play. It can also cause the unit to automatically go into fast forward or rewind; this can be disconcerting if you're adjusting or repairing the unit. If you move the VCR or move the light source above the VCR while you are working on it and all of a sudden the unit rewinds or fast forwards, you might have moved it into a position where too much light is on one of the sensors. If this happens, remove or shade the light on the VCR.

Electrical Shock

Now that you've removed the main top cover, it's time to learn more about avoiding an electrical shock. I want you to feel comfortable working on your VCR. Every VCR has a manufacturer's warning label that cautions you about electrical shock. You can touch the electrical circuits while the unit is on without getting shocked, as shown in Figure 2.8. All units run on a low DC voltage, ranging from 5 to 30 volts. This voltage isn't harmful.

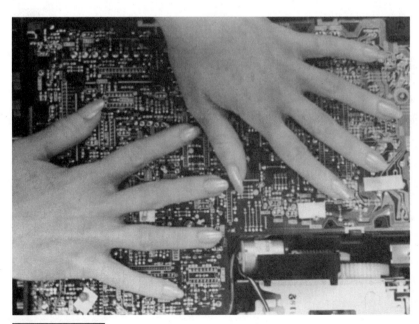

FIGURE 2.8

Touching the electrical circuits.

The only section that can harm you is the power supply. The power supply converts the 110 volts into a lower voltage. To locate the power supply, follow the power cord through the back of the unit and into a covered connection box. Inside the box is the main fuse. The connection box is usually encased with plastic or metal and it's grounded to the chassis to protect you. Do not under any circumstances remove this protective cover while the unit is plugged in! From this box, the 110 volts go directly into the power transformer, as shown in Figure 2.9. The power transformer reduces the 110 voltage to a variable low-voltage source, which makes the VCR function.

In a few models, the 110 volts is exposed, as shown in Figure 2.10. Follow the power cord through the back of the unit to where it connects to a circuit board. Beside this connection on the circuit board is the main 110 volt power fuse. Behind the power fuse is a 110-volt AC outlet. These are the only two points where you can get shocked. From here, the power goes directly into the power transformer.

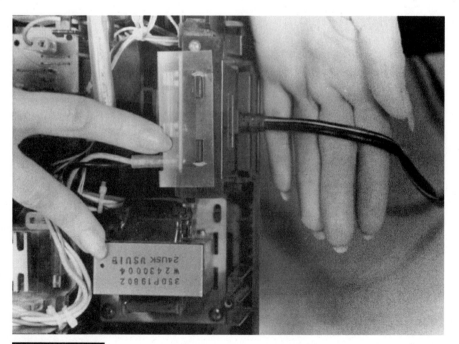

FIGURE 2.9

A covered box on the top, and power transformer on the bottom.

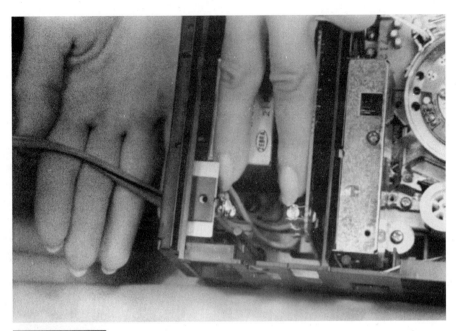

FIGURE 2.10

Exposed 110 volt connections.

Caution!

The places where you can get shocked are:

- where the power cord connects to the circuit board
- the main power fuse beside the power cord
- all components between the power cord and the power transformer.

The power transformer (Figure 2.14) is the biggest component in the power supply. It has a rectangular shape and is placed in the middle of the power-supply section, as shown in Figures 2.12, 2.13, and 2.14.

To prevent electric shock, you'll to need cover the power supply with soft plastic or cardboard. To do this, place a flat piece of cardboard over the top of the components on the circuit board from the power transformer to the power cord. Then, secure it down with black electrical tape to hold it into place, as shown in Figure 2.15.

QUICK»TIP If you find an exposed 110-volt connection, unplug the unit and place black electrical tape over the exposed area. The tape protects you against electrical shock, as shown in Figure 2.11. Another precaution is to take a thin sheet of plastic or cardboard, placing it over and around the entire exposed area. Then use black electrical tape to hold it down.

On all newer models, the VCR has a small power transformer, the voltage is reduced electronically in the power supply, the power supply is located on the motherboard. To locate the power supply, follow the power cord into the unit and follow it to the main power fuse. The power fuse is located inside the power-supply section on the motherboard. Some boards are marked with a white coating covering the entire power supply section, as shown in Figure 2.12. The rest of the board will be brown or green in color. Other power supplies will have a white line around the hot part of the power supply with the word "hot" printed inside the white line on the circuit board, as in Figure 2.13. This portion of the circuit board has 110 volts and can shock you.

Working Conditions to Avoid

1. Working on a VCR on or around an electric stove, or any other electrical apparatus.

2. Grabbing hold of the TV coax cable connector while working inside the VCR.

3. Working close to any sink made of metal.

4. Standing bare footed on a cement floor.

5. Standing in a puddle of water.

6. Wearing rings while working on a VCR.

7. Soldering a part while the unit is plugged in.

The VCR should be plugged in only when:

1. You are diagnosing the problems in the VCR by watching the lines on the TV monitor.

2. The VCR is in play, fast-forward, or rewind, and you're looking for a mechanical problem.

3. Checking a recording problem.

4. Checking the audio amplifier.

5. Checking the video head amplifier.

6. Doing an audio or video alignment.

7. Making the torque adjustment.

8. Centering the tracking control.

9. Checking for a bad DC motor.

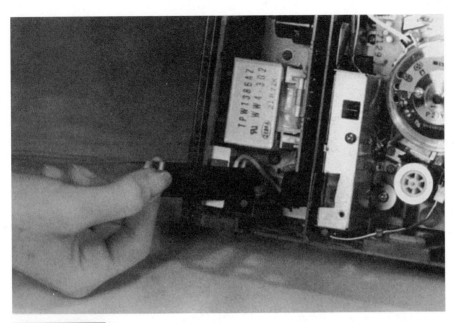

FIGURE 2.11

Covering exposed connections with electrical tape.

FIGURE 2.12

White coating on circuit board showing the power supply section.

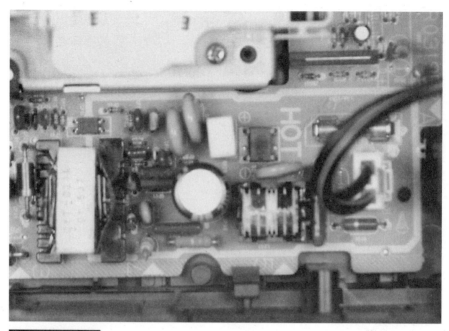

FIGURE 2.13

White line around the power supply with the word "HOT."

FIGURE 2.14

Exposed 110 volt connections from the power transformer to power cord.

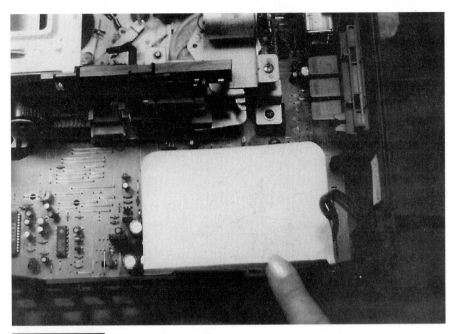

FIGURE 2.15

Covering the exposed connections with cardboard.

Under all other circumstances, the VCR should be unplugged before you begin repairs.

Review

1. Always be sure to have the proper tools.

2. If the top cover doesn't come off easily, look for another screw.

3. Cover any exposed 110-volt terminal with black electrical tape to protect you.

4. Never stick a screwdriver in the covered AC box.

5. Never lay tools on top of a circuit board.

6. Never grab an electrical appliance while working on a VCR.

7. Never grab a TV cable connector while working inside the VCR.

8. Do not wear rings while working inside the VCR.

9. These precautions are here to protect you. Follow them carefully.

I've covered many types of top covers and methods for removing them.

Conclusion

The next chapter covers the most crucial portion of a VCR: the path the video tape travels.

CHAPTER

3

Locating the Video Tape Path

Before you can learn how to clean and service a VCR, you must first learn the videotape path. The videotape path is the most crucial portion of a VCR. Hundreds of different makes and models of VCRs are available. However, from the very first model produced to the most current on the market today, the videotape path remains the same. Study Figure 3.1, then I'll explain the videotape path.

How the Tape Path Works

When a tape is loaded, the videotape path starts from the left side of the videocassette. As the tape comes off the supply reel, its first contact is the supply tape guide inside the videocassette. The supply tape guide secures the videotape on its proper path. The tape then passes over the back tension guide, which maintains the right amount of torque. The tape travels across a supply tape guide attached to the transport. The transport is the chassis of the carriage; most all of the moving and stationary parts are attached to it. The supply tape guide secures the videotape on its proper path so that the back tension guide

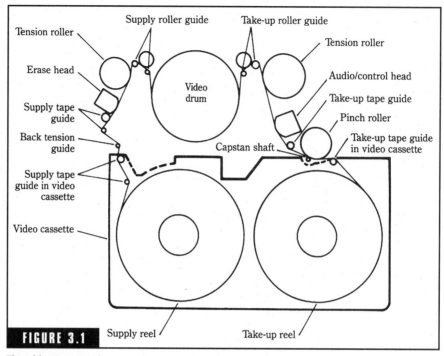

Tension roller
Supply roller guide
Take-up roller guide
Tension roller
Erase head
Video drum
Audio/control head
Supply tape guide
Take-up tape guide
Back tension guide
Pinch roller
Capstan shaft
Take-up tape guide in video cassette
Supply tape guide in video cassette
Video cassette

FIGURE 3.1
Supply reel
Take-up reel

The video tape path.

and the erase head can be aligned to the videotape properly. The tape then passes the erase head, which erases previously recorded information when in the Record mode. Next, the tape travels up against the supply tension roller. The tension roller prevents the tape from shuddering and keeps the tape running smooth. The tape travels around the supply roller guide, aligning the video tape to the left side of the video drum. The video drum houses the video heads, which read the signals off the videotape. This process enables you to see images on the TV screen.

The videotape continues around the video drum and travels to the take-up roller guide. The take-up roller guide aligns the videotape to the right side of the video drum. In some models, the tape passes up against another tension roller. It then travels to the A/C head. *A/C* represents audio and control. The A/C head's function is to play and record the audio portion and to control the sync of the image in Play or Record mode. The tape crosses the take-up tape guide, securing the videotape on its proper path so that the A/C head and the pinch roller can be aligned to the videotape properly. It then travels between the capstan shaft and pinch roller. The capstan shaft turns and pulls the videotape through the tape path. The pinch roller clamps up against the capstan shaft to provide the pressure necessary for the capstan shaft to pull the tape. The tape continues to another take-up guide in the videocassette. This guide aligns the videotape to the take-up reel, the spool that the videotape rewinds on inside the videocassette.

> **TAPE PATH NEW MODELS**   **TRADE SECRET**
>
> The only difference between the newer models and the older models is the placement of the tension rollers. The supply tension roller on the left side of the tape path has been moved from behind the erase head to the front of it, as shown in Figure 3.2. The other tension roller on the right side, the take-up tension roller, has been eliminated.

Shield Plates

In many VCRs, a shield plate is present to keep dust off the video heads and prevent electrical interference. A shield plate must be re-

FIGURE 3.2

Supply tension roller.

FIGURE 3.3

Shield plate and grounding wires.

moved to service and clean the videotape path. The shield plate is shown in Figure 3.3.

To remove the shield plate:

1. Unplug the VCR and locate the screws holding the shield on. Sometimes there is one mounting screw, but there can be up to six. There also might be a wire attached to some of the shield plate screws, (see Figure 3.3). You might find two wires attached to two plate screws, or just one wire attached to one plate screw. These attached wires are called *grounding wires* and they are used to ground the shield plate as well as other parts of the VCR.

2. If a wire is attached to a plate screw, remove the screw, pull the wire clip off, and tuck the wire or wires out of the way. Figure 3.4 shows that the wires have been tucked away to keep them from shorting out other circuits.

3. Check to see that you've found all the screws. On certain models, you might find a shield plate screw mounted to the transport, as shown in Figure 3.5. While reaching down to retrieve the shield, it's possible to drop the screw inside the unit and lose it. If this happens, a magnetic screwdriver will come in handy.

> **QUICK>>>TIP**
>
> **To make your screwdriver into a magnetic screwdriver, refer to the section in Chapter 1, "Magnetizing a Screwdriver."**

4. After removing all the mounting screws, remove the shield plate by grabbing it and pulling it straight up and out. Figure 3.6 shows the videotape path after removing the shield plate.

FIGURE 3.4

Tucking away the grounding wires.

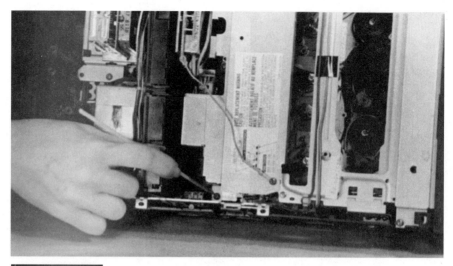

FIGURE 3.5

Shield plate screw mounted to the transport.

FIGURE 3.6

Removing the shield plate.

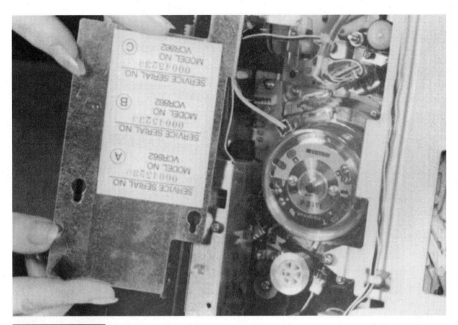

FIGURE 3.7

Removing a shield plate.

Other Considerations with Shield Plates

■ Some shield plates have no grounding wires to be removed. On this type of shield plate, you don't have to completely remove the screws. Just loosen the screws, slide the shield plate to the left, then pull the shield plate straight up and off, as shown in Figure 3.7.

■ Another type of shield plate has screws and clips holding down the cover plate. After removing the screws, you'll find two black latches in each corner on the backside of the plate. The latches are shown in Figure 3.12. Take hold of one latch, push it back, and lift up on the plate at the same time. Do both latches in the same manner.

■ Some shield plates have an extension arm going up one or both sides of the cassette carriage, as shown in Figure 3.8. You'll find one or two mounting screws on or at the end of this extension.

QUICK»TIP

Most new models have no shield plates blocking the tape path.

FIGURE 3.8

The shield plate extension arm.

Circuit Boards

Often, a circuit board will be blocking the tape path, as well. As with the shield plate, the circuit board must be removed before you can access the tape path.

Circuit Board Removal

There are four ways to remove the circuit boards in your VCR. One of them will be appropriate for your particular model.

1. In some models, a circuit board is placed over the top of the tape path and is usually hinged. The hinges are usually at the back of the board (as you face the unit). Mounting screws keep the front of the circuit board mounted to the cassette carriage. The cassette holder is the compartment where you insert the videotape. Two or three mounting screws are on this board, as shown in Figure 3.9. Remove these mounting screws.

   A few models have a grounding leaf spring. The leaf spring grounds the main top cover to the chassis. Most grounding springs

FIGURE 3.9

Locating the mounting screws on a circuit board.

are mounted with a Phillips head screw, as shown in Figure 3.10. This spring must be removed to pull up the circuit board. Remove the screw, take hold of the front of the circuit board and pull it straight up. The board opens like a door, exposing the shield plate. In Figure 3.11, this model has two mounting screws and no grounding wires on the shield plate. Remove the screws and lift off the shield plate.

2. Other models with circuit boards covering the tape path have one to three screws on the board to be removed. These screws are usually red in color and have arrows or circles around them printed on the circuit board, (refer to Figure 9.2 and Figure 9.7). Look for one to four black latches around the outside of the circuit board or protruding through a small rectangular hole in the middle of the circuit board. To release these latches, pull straight back away from the board simultaneously lifting that portion of the circuit board, (see Figure 3.12). Release each latch until the circuit board opens like a door.

3. Some models have two mounting screws near the top in each corner of the back of the unit, as shown in Figure 3.13. Remove these

FIGURE 3.10

A grounding leaf spring.

two screws. Next, you will find one latch on each side of the circuit board. To release these latches, pull straight back away from the board while simultaneously lifting on that portion of the circuit board (refer to Figure 3-12). Release both latches, lift up on the front of the circuit board, and the entire back of the unit will open like a door.

4. Some models have four to six mounting screws instead of hinges. These screws are marked with an arrow or have a circle around them. The screws might be red in color. Usually, one or two screws are attached to the cassette carriage, one or two screws are attached to the chassis, and two screws go directly through the board and attach to the back of the unit. Remove all screws and lift the circuit board straight up. In these models, part of the back comes out with it, containing the in and out video plugs, as shown in Figure 3.14. As you remove the board, flip it over the back of the unit onto the table. In other models, you can only move the board over to the side.

FIGURE 3.11

A shield plate and opening a circuit board.

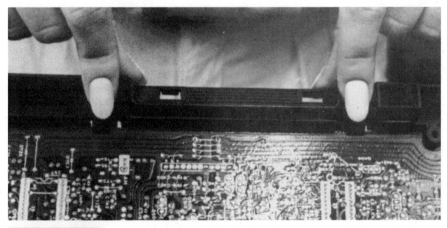

FIGURE 3.12

Releasing the latches.

FIGURE 3.13

Mounting screws on the back of the unit.

FIGURE 3.14

Removing a circuit board.

Holding the Circuit Board Open

To keep a circuit board with hinges open for cleaning and repairs, place any non-metal object (such as a glass brush) between the top of the chassis to the right side of the cassette carriage and the bottom of the circuit board.

Review

> **QUICK>>TIP**
>
> Sometimes, the two screws mounted to the back of the unit are black; they go through the plastic panel that contains the in and out video plugs or a screw night be directly beneath the in and out video plugs. For additional information on removing circuit boards, refer to the section in Chapter 31, "Boards underneath the top cover."

1. The tape path is the same in all models of VCRs, except for the placement of tension rollers.

2. In some of the older models, the tape path is covered by a shield plate and must be removed for servicing.

3. Remember to reattach any grounding wires to the shield-plate mounting screws.

4. Use a magnetic screwdriver for those hard-to-access screws.

5. All screws have different threads, so when replacing the screws, be sure that you put the right screw in the right hole.

6. Most units with circuit boards are hinged and open like a door.

7. Remove the mounting screws marked with red screws, arrows, or circles around them.

8. In some models, remove the circuit board and part or all of the back cover will come up with it.

9. All newer models have no shield plates or circuit boards to remove.

Conclusion

The next chapter covers the cleaning process. If good maintenance is followed, a VCR functions more efficiently and generally has fewer mechanical malfunctions.

The Cleaning Process

It's **important to** be prepared by having all necessary tools and materials. Be sure to gather everything needed before beginning the cleaning process.

Materials

1. One can of degreaser
2. A roll of paper towels
3. Cleaning alcohol or head cleaner
4. Chamois sticks
5. Glass brush
6. A soft-bristled paint brush
7. An empty lid or a small container

Removing Dust and Dirt

Unplug the VCR and tip the unit on its side with the cassette carriage side down. In older units, the cassette carriage will be on one side or the other. In newer units, the carriage is in the middle of the unit and can be placed on either side. If the unit has accumulated a large amount of lint and dust, the best way to blow the dust out is with an air compressor, vacuum cleaner, or a can of dust-remover spray. Otherwise, the job can be performed manually. Take a paintbrush and start dusting off the VCR at the top of the unit where the circuit board is. Work your way down to the bottom of the unit, as shown in Figure 4.1. By dusting off the VCR in

FIGURE 4.1

Dusting a VCR.

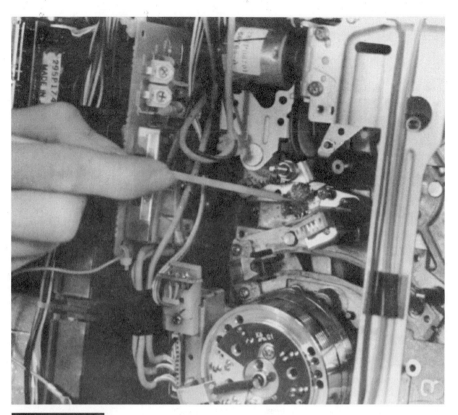

FIGURE 4.2

A sticky film caused by dust and oil.

this manner, the dirt will fall onto the table, rather than inside the unit. When you reach the video drum, be careful to brush only the top of the drum and not the sides of the drum. Brushing the sides can cause damage to the video heads. Be sure to get inside all of the nooks and crannies. VCRs collect a lot of dust.

The brush is unable to reach some areas. You might find that dust and oil have accumulated, causing a sticky film. The brush will only smear the film and not remove it, as shown in Figure 4.2. Remedy

QUICK>>>TIP

I've found instances in which a silicone spray is used to clean or lubricate. If this is used, a VCR can completely stop functioning. Silicone spray leaves a film residue that causes the belts and pulleys to slip. If this has happened, just re-spray it with a can of degreaser. The degreaser will remove all the silicone spray. After you're satisfied that the chassis is clean, lay the unit flat down on the table.

FIGURE 4.3

Removing the sticky film by spraying it with degreaser.

this by taking a can of degreaser and spraying from the top to the bottom of the cassette carriage. This rinses the residue toward the bottom end of the unit for removal.

Be sure to spray into all those hard-to-reach areas by using the nozzle extender, as shown in Figure 4.3. Spray liberally; degreaser won't damage any part of a VCR. Take a paper towel and collect the dirt that has accumulated at the bottom of the unit, as shown in Figure 4.4.

Cleaning the Various Components

After you have given the unit a general cleaning, you should give special cleaning care to some parts. These parts and the methods to clean them are described in this section. To clean the videotape path, pour head cleaner or cleaning alcohol into a small container. Take a chamois stick and immerse it in the solution.

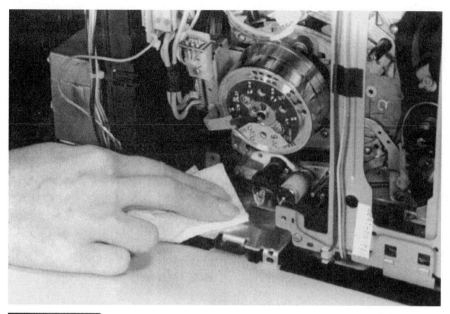

FIGURE 4.4

Removing accumulated residue.

Capstan Shaft

Figure 4.5 shows the location of the capstan shaft. A black residue will be caked on the surface of the capstan shaft. This residue needs to be removed. Take a saturated chamois stick and rub it up and down the shaft until the exposed portion of the shaft is clean. Sometimes this black residue is quite hard and a chamois stick can't remove it. To remove this residue, take a glass brush and clean it in the same manner, (refer to Figure 27.7).

In some units, the capstan shaft is easily accessible and you can clean all the way around it without turning the shaft. In newer VCRs, the front position of the capstan shaft is the only area that you can reach, as shown in Figure 4.6. You need to remove the upper portion of the pinch-roller assembly before you can reach the capstan shaft for cleaning. Proceed to the section in Chapter 7 entitled "Removing pinch rollers." After cleaning the exposed area, the next step is to rotate the shaft. Turn the shaft by placing your finger on the exposed part of the shaft and sliding your finger across the shaft. All new models have a direct drive motor and turn very easy. In other direct-drive

FIGURE 4.5

The capstan shaft.

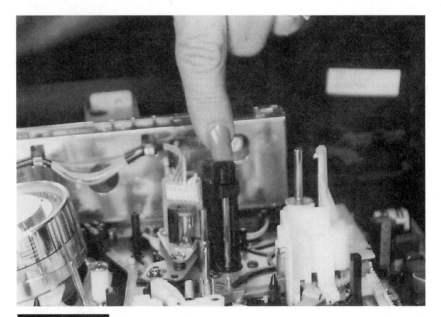

FIGURE 4.6

Partially exposed capstan shaft.

If the flywheel is not exposed, use the following procedure. Plug in the unit, insert a videotape, push play, and then stop. Eject the tape. The capstan shaft stops in a new position. Now you can clean this portion of the shaft. You might need to repeat this step two or three times to clean all the way around the shaft. When the capstan shaft is completely clean, it'll be nice and shiny.

models, you can grab the shaft and turn it. With some models, the capstan shaft doesn't turn as easily, but it does turn. If it absolutely won't turn, the unit is belt driven. Look for a capstan motor flywheel, as shown in Figure 4.7. If the unit has an exposed flywheel, place a finger on top of it and spin it in either direction. This procedure easily turns the capstan shaft for cleaning.

FIGURE 4.7

The capstan motor fly-wheel.

Pinch Roller

Place your finger on top of the pinch roller. Because the pinch roller moves freely, a finger will stabilize it, as shown in Figure 4.8. In newer models, the pinch roller is mounted upside down, but you can still place your finger on top of the pinch roller to stabilize. Take a saturated chamois stick and rub it up and down the roller until it's clean. Using your finger, rotate it slowly until you've cleaned all the way around it. Notice that a lot of black residue will come off the pinch roller. You might need to use two or three different chamois sticks to completely clean the pinch roller.

A/C Head and Eraser Head

Figure 4.9 shows the location of the A/C head. Place a clean chamois stick horizontally across the A/C head and rub it back and forth until

FIGURE 4.8

Cleaning, and stabilizing a pinch roller.

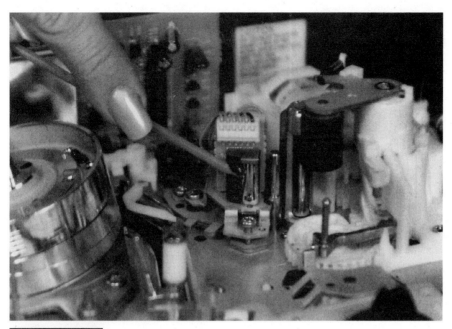

FIGURE 4.9

The A/C head.

the head is clean. Sometimes you'll find more hard black residue buildup on the A/C head. This residue is hard to get off. Just keep scrubbing until it's removed or use a glass brush to remove it. If you're using a glass brush, use a chamois stick to clean any fibers left by the brush. Do not try using a screwdriver or a hard object to remove these hard spots; you'll damage the A/C head! Locate the eraser head and clean it in the same manner as you cleaned the A/C head.

Tension Rollers

The next part to clean is the tension roller. Some models have two tension rollers, one on each side of the video drum. Other models contain one tension roller to the left side of the video drum (refer to Figure 3.2), and a few models have none. You can clean the tension rollers in the same manner as cleaning the pinch roller. As you follow the video tape path around, the next part you will come across is the video drum. Leave it for now. I'll show you how to clean the video drum in Chapter 5. For now, continue on to the next tension roller.

Roller Guides

Place a chamois stick in a horizontal position onto the glass portion of the roller guides. Scrub up and down to remove any residue. Be sure to go all the way around each roller guide (refer to Figure 12.3).

Review

1. Tip the VCR on its side with the cassette carriage side down. If the model has the carriage in the middle, tip it either side.

2. If you have an air source, use it to blow out all the dust. If you don't have an air source, brush off all dust, starting at the top of the unit and working down to the bottom.

3. If necessary, spray it with degreaser.

4. Lay the unit flat on a table, then clean the video tape path using a saturated chamois stick.

5. Be sure to clean:

   A. Capstan shaft

   B. Pinch roller

   C. A/C head

   D. Tension rollers

   E. Eraser head

   F. Roller guides

QUICK»»TIP **Make a checklist of parts to help you remember what parts you've cleaned and what parts still need to be cleaned.**

Cleaning the Video Heads

All **video heads** look and work the same way. The video heads are located inside the video drum. I've removed a video drum from a VCR to demonstrate its functions.

The Video Heads

A video drum consists of two parts:

1. The upper video drum that rotates and

2. The lower video drum that is stationary and is bolted to the transport.

As you observe the upper video drum, you'll notice a little slot on each side, directly across from each other, as shown in Figure 5.1. These slots contain the video heads. In each slot is the protruding tip of a video head. This tip is what you need to clean.

I've disassembled the upper video drum to give a better view of the video heads (Figure 5.2). This particular model has two heads. In

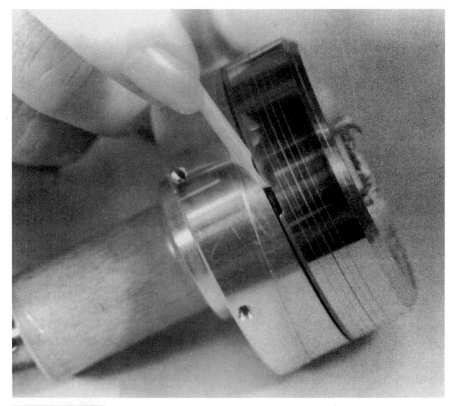

FIGURE 5.1

The slots in an upper video drum.

some models, you might find three or even four video heads or slots on the side of the video drum.

Caution

When working on a VCR, do not, under any circumstances, place your fingers over the slots on the video drum. This could cause damage to the video heads. The video drum is very delicate and expensive.

Why Video Heads Stop Working

Look at a close-up view of a video head, as shown in Figure 5.3. The video heads are made of a metallic carbon compound, which is extremely fragile and expensive. Notice the view port or gap of the video

FIGURE 5.2

The video heads.

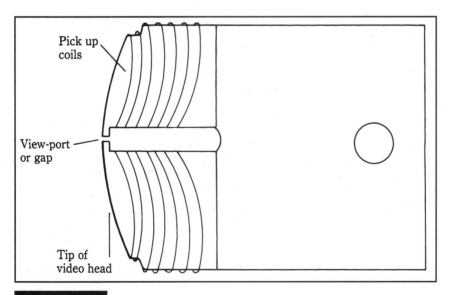

Pick up
coils

View-port
or gap

Tip of
video head

FIGURE 5.3

A close-up view of a video head.

head. This portion reads the signal off the videotape. One reason that a video head malfunctions is that the view port is obstructed with dust or lint from dirty tapes. Consequently, the head cannot read the signal. The video heads are quite active—the heartbeat of the VCR. Keeping them clean is extremely important.

Cleaning the Heads

To clean the heads:

1. Pour some head cleaner or cleaning alcohol into a small container.

2. Take a new chamois stick and saturate it in cleaner.

3. Apply the chamois stick, as shown in Figure 5.4, in a flat position so that half the chamois is on the lower drum and half is on the upper drum. Hold the stick at a slight angle.

4. Place your finger on top of the video drum.

5. Start rotating the upper drum in a counterclockwise direction, rotating one revolution every four seconds. As you slowly rotate the

FIGURE 5.4

Applying a chamois stock at the correct angle, and rotating the upper video drum.

FIGURE 5.5

Moving the chamois stick up, and off the video drum.

upper drum, move the chamois stick up and off the drum in one direction, toward the top of the video drum, as shown in Figure 5.5. Do not pull the chamois stick away from the video drum before you reach the top. Let the chamois stick go all the way up and off. This way you won't leave any dirt deposits on the drum.

Repeat this procedure two or three times.

Caution! Using a chamois stick in an up and down motion causes damage to the heads. Do not use this motion (Figure 5.6).

In some models, a rotary cap is on top of the video drum. A rotary cap is the part of the drum motor that makes the upper drum spin. Other models have an exposed circuit board or what is called a *stator* (see Figures. 5.7 and 18.14). In either case, place your finger on the top corner edge to rotate the upper drum, as shown in Figure 5.7. Clean the video drum (as explained previously). Avoid using cotton swabs because their fibers get lodged inside the view port making the head dirtier.

FIGURE 5.6

An incorrect position for a chamois stick.

HEAD CLEANING TAPES

Cleaning tapes don't always work. Some cleaning tapes just smear the dirt and film, leaving a trail of dirt deposits in a ring around the upper video drum. Smearing the dirt deposits can clog the view port of the video head even more (refer to Figure 5.8). The construction of the video head creates a need for a certain amount of pressure to be applied to remove the dirt from the view port. This pressure is applied when cleaning the video drum by hand. Keep the chamois stick flat against the video drum at a slight angle and apply a steady amount of pressure as you move the chamois stick up and off the drum.

Built-in Head Cleaner

New VCRs have a built-in head cleaning pad (refer to Figure 5.9). This pad momentarily contacts the video drum as the videotape is being loaded onto the tape path, brushing off any loose dirt before the videotape contacts the video drum. If this cleaning pad is dirty and a black residue ring is around the upper drum (as shown in Figure 5.8), you need to change the pad. If the cleaning pad is made of foam rubber (Figure 5.9), carefully pull the pad straight up and off the plastic shaft. This should come off easily. To change non-foam rubber cleaning pads, look for a flat plastic cap on top of the cleaning pad. Place a flat-head screwdriver at the base of the cleaning pad as-

FIGURE 5.7

Rotating a video drum with a rotary cap.

FIGURE 5.8

Residue ring around the upper video drum.

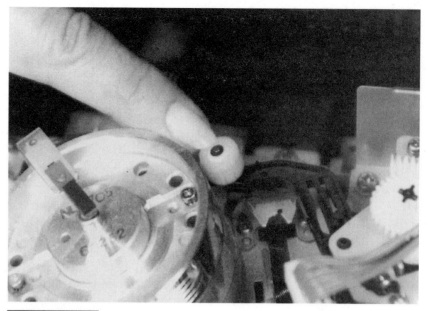

FIGURE 5.9

Head cleaning pad.

QUICK≫TIP

CONNECTING THE VCR TO A TV MONITOR

1. Attach a coax cable to the antenna out plug of the VCR and the other end to the antenna plug on the TV.

2. Attach the antenna, or outside cable, to the antenna in plug on the VCR.

3. Plug in both units and turn them on.

4. Select the proper channel, 3 or 4, on the TV. The VCR/TV switch must be placed in the VCR position.

5. Select a channel on the VCR with a broadcasted picture. This shows the VCR is producing the signal, not the TV. If both units are connected and fine tuned properly, you should have a picture of what is being shown on that channel.

sembly and apply some pressure downward so that the plastic mount won't break, as shown in Figure 5.10. At the same time, pull up on the cap on top of the pad to remove it. Once the cap is removed, pull the pad off. Be careful because some pads consist of hundreds of tiny fiber washers that can fly all over the place. To purchase a new cleaning pad, refer to the "Parts" section in Chapter 1.

Now that the video heads and the entire unit are clean, it's time to hook up the VCR to the TV monitor and learn its functions.

By following this procedure, you've turned the TV into a monitor, a useful tool for diagnosing problems in a VCR.

FIGURE 5.10

The cap on top of the pad and pushing down on the arm of the bracket.

Dirty Video Heads

Some VCRs have two video heads, and others have four. The symptoms of dirty heads are different for each, as are the steps necessary to get the heads working properly again. See the following for a section that pertains to your VCR for cleaning instructions.

A TWO-HEADED MACHINE

Now that everything is ready, insert a video tape and push play. The VCR should be producing a good picture. Now I'm going to tell you the symptoms of a dirty video head and describe what you'll see on a TV monitor.

A FOUR-HEADED MACHINE

What happens if one of the four heads is clogged and you insert a two-hour video tape and push Play? The image on the TV screen is okay, but when you insert a six-hour video tape, the image is distorted or

◈TROUBLE-SHOOTING

SYMPTOMS

If the heads are partially clogged, you'll receive a partially distorted picture, as shown in Figure 5.11. You'll probably experience a loss of color as well.

When both video heads are badly clogged, the entire picture is completely snowy.

There are a few VCR's that mute the video signal and cause the TV screen to go completely black or blue whenever a tape is inserted into the machine. If the audio portion of the video tape sounds okay, usually the cause is a clogged video head.

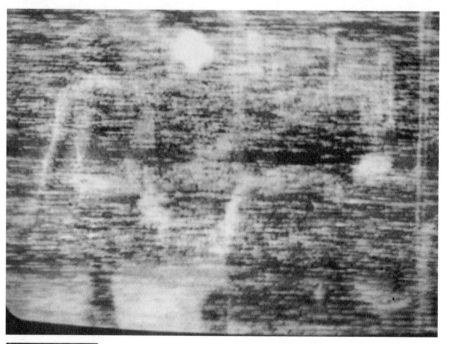

FIGURE 5.11

A partially distorted picture.

snowy. On the other hand, the two-hour tape could look snowy or distorted, but the six-hour tape looks okay. What I'm saying is this: two of the video heads affect the six- and four-hour video tapes, while the other two heads affect the two-hour video tapes. To correct this problem, you'll need to clean all four heads, as previously explained. After cleaning the video heads, if you are still receiving a distorted picture, clean it again. I've been able to save many hours of frustration and much money by recleaning the video heads.

Review

1. Remember that all video drums look the same, although some have a rotary cap on top.

2. The drum has two parts:
    A. The upper video drum holds the video heads and rotates.
    B. The lower video drum is stationary and supports the upper drum.

3. Never remove or disassemble the video drum unless you are going to replace the video heads.

4. You need to clean the upper drum and the tips of the video heads.

5. Saturate a clean chamois stick with cleaning solution.

6. Move the chamois stick up and off the video drum in one motion, at a slight angle, not up and down.

Conclusion

This chapter has given you a lot of information to absorb. You should be feeling comfortable with the information and your capabilities. The next chapter is about "Saving a video head." Cleaning the video heads, as well as the other routine cleaning procedures, comes down to proper maintenance.

Pinpointing and Correcting Video Head Problems

If you properly clean the video heads and still have no picture, the next step is to determine if the VCR has a bad video head or a circuit problem. Refer to the section in Chapter 5, "Connecting the VCR to a TV monitor."

Detecting a Circuit Problem

Find the TV-Video switch on the face of the VCR and place it in the Video position. Then tune in a broadcasted channel using the tuner on the VCR. If you can properly fine-tune a channel, 90% of the unit is working correctly. You've eliminated all circuit problems, except for the video head amplifier and the video heads. The next step is to determine if the problem is in the video head amplifier.

Finding the Video Leads in Older Units

Locate the video leads to check the video amplifier. Usually, two to four large, round leads run from the bottom rear portion of the lower

A few years after those models were made, the manufacturers started using a video plug instead of soldering the video leads to a small circuit board. In these models, two large, round leads protrude from a white plug mounted at the rear of the lower video drum.

video drum. The shielded cable houses the video leads. Leads are usually gray or white in color, but in a few models, they are red or yellow.

Old models have two large, round leads soldered to a small circuit board, mounted right behind and on the lower video drum. (See Figure 6.1. I've removed the video drum to give you a better view.) In a few models, the circuit board is covered with a small shield plate, held on with one screw. You'll need to remove this plate.

In some models, the white plug is at the right rear portion of the video drum, as shown in Figure 6.2. In other models, the plug is horizontally placed right behind the video drum.

In stereo hi-fi models, you might find three large, round leads coming from a white video plug at the rear of the lower video drum, as

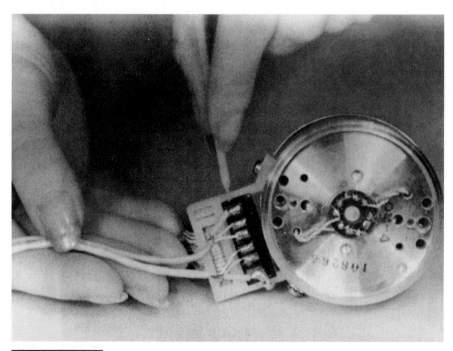

FIGURE 6.1

Locating the video leads and connections.

FIGURE 6.2

Locating the video plug.

shown in Figure 6.3. Follow the leads coming from the video plug. Two of the leads pair up and go to a small, covered metal box to the right of the video drum, usually located near the top of the unit. These two leads are the video leads. The third lead runs to a different circuit

FIGURE 6.3

Locating the three shielded video leads, in stereo hi-fi models.

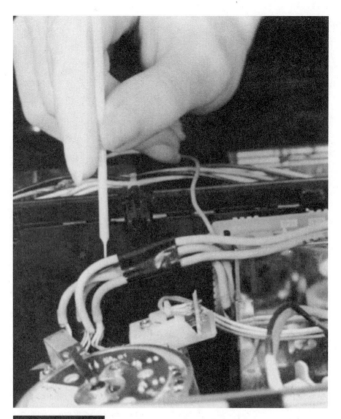

FIGURE 6.4

Locating the four shielded video leads, in stereo hi-fi models.

board, usually near the bottom of the unit. This lead is for the stereo hi-fi, or audio section.

In other stereo hi-fi units, you'll find four large round leads coming from the video plug as shown in Figure 6.4. Follow these leads. They pair off and go into two different sections of the unit. One pair of leads goes to a covered metal box to the right of the video drum, usually near the top of the unit. These leads are the video leads. The other pair of leads usually go to a circuit board, near the bottom of the unit. These leads are the audio leads.

A few models have a rotary cap on top of the upper video drum. One or two white plugs are connected to a circuit board attached to the back of the rotary cap. Do not confuse these two plugs with the video plug. The plug to the right of the lower video drum, as shown in

FIGURE 6.5

The rotary cap plugs on top, and the video plug on the side.

Figure 6.5, is the video plug and has only two large round leads coming off it. This is the plug you're looking for.

Finding the Video Leads in Newer Units

Instead of using large round leads newer units now use a ribbon. A ribbon is a group of wires inside one insulation and is flat in appearance. The ribbon protrudes out from behind the lower video drum. At the opposite end of this ribbon are exposed bare contacts, where the ribbon is folded back, as shown in Figure 6.6. These are the video leads, or connections. In most models, the ribbon goes into a covered metal box or the video head amplifier right behind the video drum. This ribbon is connected to a plug in front of the video head amplifier; the exposed connections are on the ribbon right at the plug, as shown in Figure 6.7. In other models, the ribbon goes straight into a video head plug and the exposed connections are at the rear of the plug, as shown in Figure 6.8. Models with the rotary cap on top of the upper

FIGURE 6.6

Bare contacts on the ribbon.

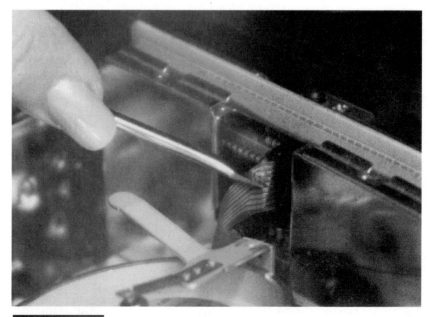

FIGURE 6.7

Bare contacts on the ribbon at the video plug.

FIGURE 6.8

Exposed contacts on a plug.

drum will have two ribbons coming out from the video drum. The ribbon you are concerned about is coming out from the lower drum. In some models, the exposed connections are on the back of the circuit board, as shown in Figure 6.9.

Checking the Video Head Amplifier

Now that you've located the video leads in the VCR, start with units using a video plug, whether or not you have a stereo hi-fi. At the end of each large round lead, just before it goes into the video plug, are three small wires connected to each large lead, (refer to Figures 6.2 and 6.3). You'll find either a blue or yellow grounding wire. You'll also see a red wire and white wire. These two are the video wires. All three wires are attached to small pins inside the video plug. The pins connected to the red and white wires are what we're concerned with.

Be sure that the TV-Video switch is in the Video position. Insert a video tape and Press play. You should see snow on the TV screen, as

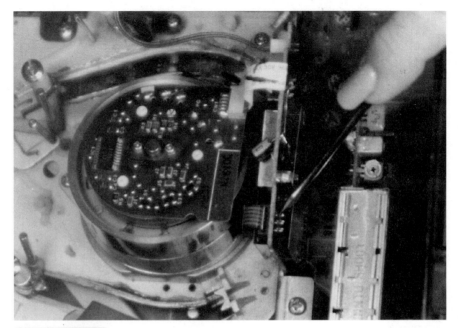

FIGURE 6.9

Contacts on video head amplifier circuit board.

shown in Figure 6.10. Find the point where the red and white wires enter the white plug and connect to the pins. Press a finger against the metal shaft of a small flathead screwdriver. Touch the corner of the tip of the screwdriver to each metal pin, one at a time. Touch the red and white wire pins, as shown in Figure 6.11. If you're not sure which pins are the video pins, touch all the pins on the video plug one at a time, starting at one end of the plug and working across it. All pins on the video plug have no voltage and will not shock you.

Keep an eye on the TV screen. Each time you touch a red or white wire connected to the pins that are directly connected to the input of the video amplifier, the snow (Figure 6.10) on the TV screen changes to many black bars or a wide black band, as shown in Figure 6.12.

In other units, you might have white bars instead of black bars, or the snow changes from a light normal snow (Figure 6.10) to a heavy darker snow (Figure 6.13). Each time you touch a video amplifier pin, a definite change is on the TV monitor. When the screwdriver is on a video pin and you touch the screwdriver, you become an antenna,

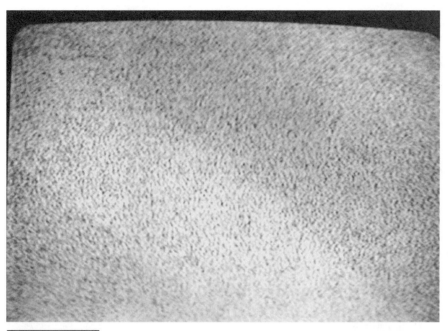

FIGURE 6.10

A snowy screen.

FIGURE 6.11

Touching the pins in a video plug.

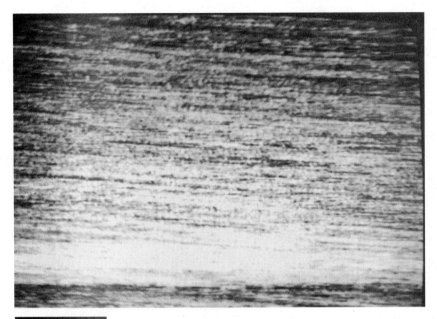

FIGURE 6.12

A black band appearing on the screen.

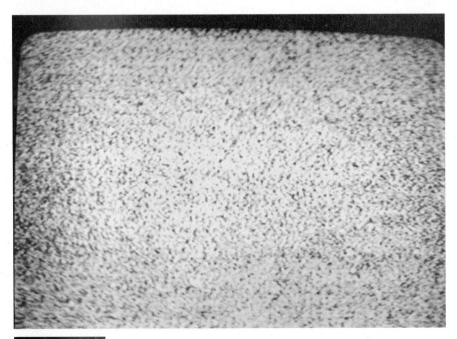

FIGURE 6.13

Heavy snow.

causing lines to appear on the TV screen. Remember that only two of the pins on the video plug change the image on the screen. All other pins on the plug have no effect on the image. Also, be careful not to short any two pins together.

If you're working on an old unit, the red and white wires are soldered to a small circuit board behind the lower video drum. Touch a screwdriver to the circuit board where the red and white wires are connected, one connection at a time (Figure 6.14). With this model, both red and white wires from each video lead changes the image on the TV monitor.

If you have a ribbon-type unit, touch each bare, exposed connection, one at a time. Only two connections on the ribbon change the image on the TV monitor. Figures 6.6, 6.7, 6.8, and 6.9 shows the location of these contacts. None of the connections carry voltage, so they will not shock you.

Some units have a ribbon that goes into a covered metal box or video head amplifier and have no bare exposed connection on the rib-

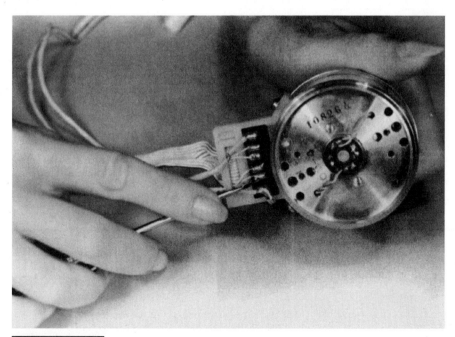

FIGURE 6.14

Touching the video contacts.

bon because it is pushed up into the plug. You have two options, either pull the ribbon out of the plug just a little bit (so that you can see the exposed area on the ribbon), or right beside where the ribbon plugs into the amplifier you can see very small pins on the side of the plug. Touch each pin on the side of the plug.

If you found at least two wires or two pins from each video lead that changed the image on the TV monitor, the video amplifier is functioning. You've also eliminated all circuit problems in the VCR. Now you know the problem is in the video heads.

Take a real close look at each video head. If you have a magnifying glass, use it. Check the tip of each video head protruding through the slot. See if it might be cracked, chipped, or broken off. Compare all the video heads on the upper drum to see if they appear exactly the same. If not, refer to Chapter 18, "Replacing the video heads." If they do, the problem is a badly clogged video head or heads. Taking the unit into a service center at this time could be quite costly. The services personnel wouldn't go any further with this problem, other than to replace the upper video drum or heads. There is another option.

Saving the Video Heads

Follow the normal procedure of cleaning the video heads. Use a chamois stick to clean it at least five or six different times. If you still have no picture after cleaning the head with the chamois stick, it's time to try something new.

You need a fiberglass brush (Refer to the section in Chapter 1, "Making a glass brush.") The fibers on the brush penetrate the view port and pull out the dirt. Use a chamois stick to clean only the outside of the video drum and the tip of each video head or view port.

Before using the fiberglass brush, be sure to clean it by dipping it into the cleaning solution. Take a paper towel and wipe off any dirt. The paper towel also removes any loose fibers left on the glass brush. Clean the brush several times, as shown in Figure 6.15.

Face the front of the VCR toward you. Go to the upper video drum and rotate it until one of the video heads is facing toward the front right side of the video drum. Hold the brush at a 45-degree angle and place the brush directly behind the view port and into the slot (Figure 6.16). Move the brush forward, across the view port, staying in the

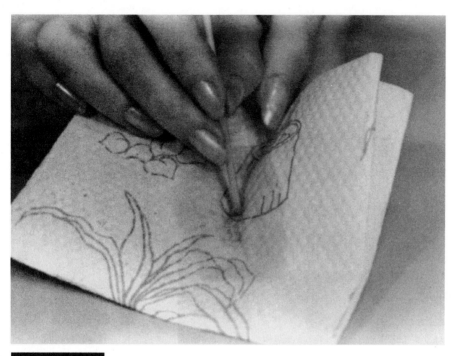

Cleaning the glass brush.

slot. Then, stop and pull the brush away from the slot. Repeat this again, staying in the slot, carefully moving the brush in only one direction across the view port. Repeat this procedure 20 or 30 times.

Rotate the upper drum and clean the next video head. Be sure to clean all video heads on the upper drum. Do not let the brush rub on the drum itself or you'll scratch it. Do not work the brush in an up and down motion or you will break the video head. When you've finished, take a saturated chamois stick and clean the upper video drum to remove any loose fibers left on the drum. (Refer to the section in Chapter 5, "Cleaning the heads.")

You are ready to insert a video cassette. Press Play to see if you have a picture. If not, repeat the cleaning procedure several more times. If this procedure fails to work, there is yet another method, however, you must be careful. Instead of sweeping the video head, as previously stated, use a scrubbing motion, back and forth only. I've found that this works as long as the video head has not been damaged.

FIGURE 6.16

Placing the glass brush in the slot.

Don't give up; eventually you'll get a partial picture, as shown in Figure 6.17. This is because particles still remain in the view port. When you reach this point, do not clean it again. Let your VCR play for an hour or so. The video head will clean itself out.

Cautions Regarding the Video Drum

There are three main concerns when you're working on or around the video drum.

1. Try not to touch the sides of the drum. You can leave oil smudges on the drum that can clog the video heads. If you have to touch the drum, be sure to clean it again.

2. When working around the video drum while the unit is running, never touch the drum with anything.

3. When you place the VCR in the Stop mode, be sure to let the upper drum stop spinning completely before you proceed to work on the VCR.

FIGURE 6.17

A partial picture.

Review

1. First, clean the video heads five or six times, using a chamois stick.

2. You need to eliminate the video amplifier. Place a finger on the shaft of a small screwdriver and touch the tip of the screwdriver as follows:

   A. To the pins on a video plug.

   B. To the connection on the circuit board directly behind the lower video drum.

   C. Where the ribbon is bare and connects to pins in a video plug.

   D. Where a ribbon is folded back and soldered.

3. After finding a video lead, look at the TV monitor to see if the image on the screen changes. If so, the video amplifier is functioning properly.

4. Clean the video heads by carefully using the glass brush.

5. After using the glass brush, don't forget to reclean the video drum with a saturated chamois stick.

6. After you finally receive a partial picture on the TV screen, do not clean the video heads again. You could make them worse. Let the VCR play and let it run until it cleans itself out.

7. Don't give up! You can save the video heads.

Servicing and Lubricating VCRs

f a VCR is a few years old or has been used for many hours, it's time to lubricate the unit—especially if degreaser has been used to clean the transport. Degreaser removes all the oil in the transport and leaves the lubricating points dry. Use a high grade of household oil to lubricate. Do not use a spray lubrication; it leaves an oily film on the belts and pulleys that causes slippage. Do not get any oil on the video drum or on any part that comes into direct contact with the video tape.

Finding the Lubricating Points

Plug in the unit, insert a video tape, and push Play. Keep an eye on the transport. As the video tape loads onto its tape path, you'll be able to see how all the moving parts operate. Stop and start the unit two or three times to find all of the lubricating points. Then, remove the video cassette.

In older units, the moving parts are interconnected, usually by a hinge. A C-, E-, or O-ring is at each hinge. Lubricate them by placing a drop of household oil on top and in the split of the C- or E-ring, as

FIGURE 7.1

A lubricating point.

QUICK»TIP To find the lubricating points under the cassette carriage, use your finger to wiggle each part back and forth, one at a time. This will show you where the parts are connected or hinged. Place a small drop of oil onto each hinge. Remember that each unit has a different combination of parts inside, but the types of hinges remain the same.

shown in Figure 7.1. To lubricate under an O-ring, place a small flathead screwdriver under one side of the split in the O-ring and pry that side up, (refer to Figure 7.4). Get a drop of oil under it and slide the O-ring back down.

In new units, the moving parts are hinged where you see a black or white round plastic shaft protruding up through the moving part, as shown in Figure 7.2. Put a small drop of oil onto each shaft or hinge where the parts inter-

Different type of lubricating point.

connect. Lubricate all hinges. Do not lubricate the video drum or get any oil on the pulleys or belts.

Lubricate the shaft on the idler wheel, pinch roller, and tension rollers, as shown in Figure 7.3. The shaft is a metal pin that protrudes through the middle of the wheel or roller. Inside the hole on top of all pinch rollers is a small Phillips screw. Some rollers have a cap on top. For removal, proceed to the section in this chapter on "Removing a pinch roller." Place a drop of oil around the screw. On top of all other shafts is a C-, E-, or O-ring. Lubricate these shafts. After lubricating each shaft, spin each roller. The pinch roller should spin freely for a second or two. The idler wheel and tension roller should turn easily, but not spin like the pinch roller. If a wheel or roller is hard to turn and won't spin freely, you'll need to pull the wheel or roller off its shaft.

QUICK»»TIP

To remove a wheel or roller, first remove the C-, E-, or O-ring connected to the top of the shaft. After removing the ring, pull the wheel or roller straight up and off. Be careful because a small washer is on top and/or under the wheel or roller. To remove rings or pinch rollers, proceed to the appropriate section in this chapter.

The idler wheel (bottom), pinch roller (center) and the tension roller (top).

Lubricate the capstan shaft. All capstan shafts have a washer at the base. Slide the washer up the shaft enough to get a drop of oil under it, (refer to Figure 25.6), then slide the washer back down. This will keep the oil under the washer and prevent it from going up the shaft and getting on the video tape. A few units have an E-ring over the washer. Remove this first (refer to Figure 25.7).

Lubricate the arm on the idler wheel assembly. Most idler arm assemblies are held on by an O-ring. Place a small flathead screwdriver under one side of the split in the O-ring and pry that side up, as shown in Figure 7.4. Then, place a drop of oil under the O-ring. Lubricate the arm on the idler wheel assemblies with C- or O-rings only.

Removing C-rings

Removing C-rings is more difficult because tension is on the ring. You have two options for removal: To purchase a C-ring puller, refer to the section in Chapter 1, "Parts," or use a flathead screwdriver to pry up

FIGURE 7.4

O-ring on the idler arm.

the ring, as shown in Figure 7.5. To replace the C-ring, place the ring on top of the shaft holding one end down with a finger. With a flathead screwdriver, pry one side of the split in the ring down over the shaft (Figure 7.6). Place the screwdriver over the other side of the split in the ring and push as before, placing the whole ring onto the shaft. Placing the screwdriver onto both sides of the split end, along with a finger, push the C-ring down the shaft until the wheel or roller is secured.

Removing E-rings

To remove an E-ring, place a small flathead screwdriver in the space between the ring and the shaft, as shown in Figure 7.7. Place your finger behind the ring so that when the ring comes off the shaft, it won't fly across the room. Twist the screwdriver and the ring will pop right off. After removing the ring, pull the roller or wheel straight up and off the shaft. You might find a small washer on the top and bottom of each roller or wheel. Be sure not to lose these washers and to put them back on the

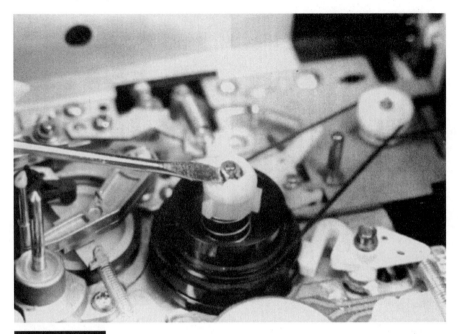

FIGURE 7.5

Removing a C-ring.

FIGURE 7.6

Remounting a C-ring.

FIGURE 7.7

Removing an E-ring off the shaft.

shaft when reassembling. To put the E-ring back onto the shaft, place the open part of the E-ring into the notch in the shaft. With a pair of long-nose pliers, snap the E-ring onto the shaft, as shown in Figure 7.8.

Removing O-rings

Place a small flathead screwdriver under one side of the split in the O-ring and pry that side up, as shown in Figure 7.9. Next, place your finger on top of the O-ring where the screwdriver is and leave it there. Pull the O-ring up and off with the aid of the screwdriver. Using the screwdriver will keep the O-ring from flying across the room. Be sure to put the ring in a safe place for remounting. To remount the O-ring, push one side of it over the top of the shaft and down into its groove with the aid of a screwdriver. Then, place your finger on the section that you just put in the groove and use a screwdriver (or your finger) and push the rest of the O-ring down into the groove, as shown in Figure 7.10.

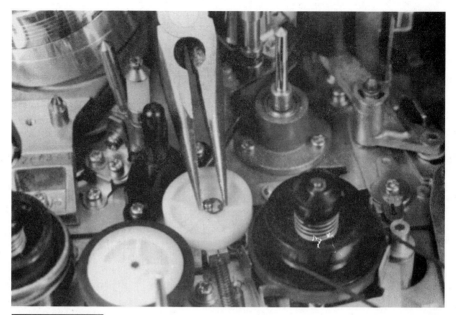

FIGURE 7.8

Remounting an E-ring onto the shaft.

FIGURE 7.9

Removing an O-ring.

FIGURE 7.10

Remounting an O-ring.

Removing Pinch Rollers

In older models, to remove the pinch roller, place a Phillips head screwdriver into the top hole and remove the screw. The location of the screw is shown in Figure 7.11. Pull the pinch roller straight up and off the shaft.

In other models, the pinch roller has a white or black cap on top. To remove the cap, place a small flathead screwdriver between the cap and the top of the pinch roller. Twist the screwdriver and the cap pops right off, as shown in Figure 7.12. Remove the screw through the hole at the top of the pinch roller.

Note the position of the pinch roller in comparison to the capstan shaft before removing the upper pinch-roller assembly. Some models have an O-ring on top of the pinch-roller assembly, as shown in Figure 7.13.

TRADE SECRET

In newer models, you need to remove the upper portion of the pinch roller assembly because the pinch roller is mounted upside down.

FIGURE 7.11

Locating the pinch roller mounting screw.

FIGURE 7.12

Removing the white cap on top of the pinch roller.

FIGURE 7.13

O-ring on top of the pinch roller assembly.

1. Remove the O-ring, grab the top of the bracket, and pull it straight up and off its shaft. In other models, look for a latch on the right top side of the pinch-roller assembly, as shown in Figure 7.14. Or, the latch could be along the side of the pinch-roller assembly, as shown in Figure 7.15.

2. In either case, pull the latch straight back to clear the assembly; at the same time, grab the top of the assembly and pull it straight up and off its mounting shaft. The top portion of the assembly and the pinch roller will pull off.

3. With one hand, grab the pinch-roller assembly. With a small flat-head screwdriver, pry the pinch roller up the shaft, as shown in Figure 7.16.

4. Pull the two pieces apart and the pinch roller and a plastic pin will come right off the shaft.

To remount the pinch roller:

1. Place the pinch roller onto its shaft and push the round plastic pin down over the shaft, as shown in Figure 7.17.

FIGURE 7.14

Latch on top of pinch roller bracket.

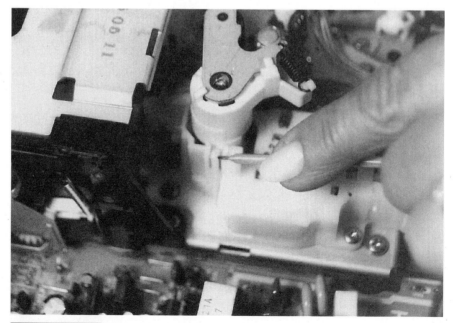

FIGURE 7.15

Latch on side of pinch roller bracket.

FIGURE 7.16

Removing the pinch roller from its shaft.

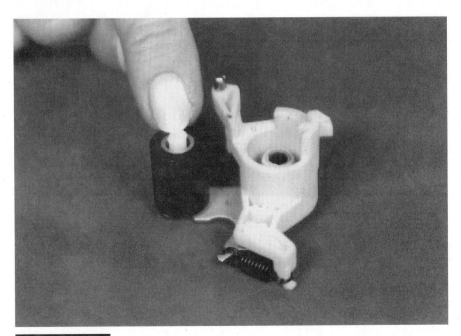

FIGURE 7.17

The pin on top of the pinch roller.

2. Place the mounting hole in the pinch-roller assembly over the shaft and be sure that the position of the pinch roller is in front of the capstan shaft. All assemblies have a pin or arm protruding out from the assembly, so all assemblies can only fit on the shaft one way.

3. Push down on the assembly so that it will slide down the shaft until it latches.

Some models have a plastic pin inserted inside the pinch roller. To remove this type of pinch roller:

1. Place a flathead screwdriver on top of the metal arm attached to the pinch roller, as shown in Figure 7.18.

2. Apply a small amount of pressure on the arm, downward; at the same time, grab the pinch roller and pull it up and off the shaft. By doing it this way, you won't bend the pinch roller arm.

3. Keep a finger on top of the pinch roller so that the plastic pin doesn't fly across the room when pulling the pinch roller off.

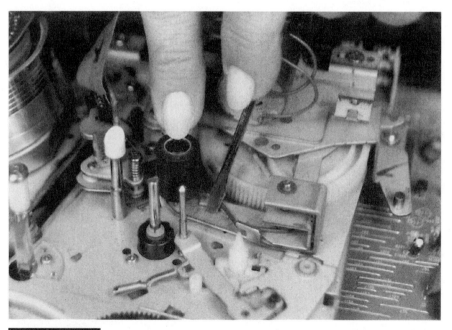

FIGURE 7.18

Pushing down on the arm and the pin inside the pinch roller.

To replace the pinch roller, place a small flathead screwdriver between the metal arm and the transport so that the arm won't bend. Place the pinch roller on the shaft and push the plastic pin down the shaft until it is tight.

Other models have a plastic or metal housing above the pinch roller, which blocks the path to the pinch-roller assembly. Sometimes three or four screws are inserted through the top of the housing. Remove the screws, or (as in this model), remove the two O-rings, as shown in Figure 7.19, and pull the housing straight up and off. Then remove the plastic pin on the shaft above the pinch roller, as explained above. If any plastic gears are on or under the housing, do not touch them, they will stay where they are. When replacing the housing, be sure that all the shafts protrude through the holes in the hous-

FIGURE 7.19

Locating the two O-rings on the housing.

ing and that the teeth of any gears are slid back together. Refer to the section in Chapter 15, "Pinch roller alignment."

Wheels That Do Not Turn Freely

For demonstration purposes, I'm using the pinch roller as a part that doesn't spin freely. After removing the pinch roller, you'll find a build-up of black residue on the pinch roller shaft. For any rollers or wheels that don't turn freely:

1. With a can of degreaser, spray off the residue onto a paper towel, as shown in Figure 7.20.
2. Wipe off any excess residue with a clean paper towel.
3. Clean out the inside of the pinch roller.
4. With degreaser, spray down inside of the pinch roller.
5. With a glass brush, spray some degreaser onto it. Use the brush to scrub the inside of the pinch roller until it's free of all residue, as shown in Figure 7.21.

FIGURE 7.20

Removing residue from the pinch roller shaft.

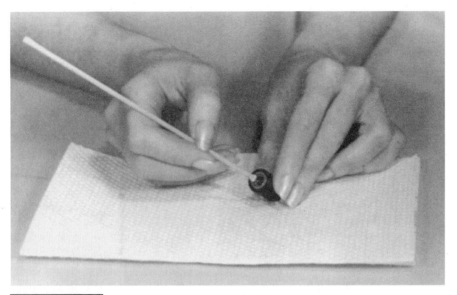

FIGURE 7.21

Using a glass brush to clean inside the pinch roller.

6. Spray the inside of the roller again to remove all fibers left by the brush.

7. Place a drop or two of oil onto the pinch roller shaft, then place the pinch roller back onto the shaft.

8. Replace the screw and cap.

The Roller Guide Tracks

Some parts in the unit need to be greased instead of oiled. Underneath the roller guides are tracks, as shown in Figure 7.22. Re-lubricate these tracks if you used degreaser on them or if the grease has been rubbed off by the roller guides. If the old grease becomes black and hard, you'll need to remove it:

1. Remove all black grease with a paper towel or spray it off with degreaser.

2. Any caked-on grease spots left on the tracks can be removed by using a flathead screwdriver wrapped in a paper towel, as shown in Figure 7.23. In some older models, push the supply tension roller over to reach the roller guide tracks (refer to Figure 7.22).

FIGURE 7.22

Location of the roller guide tracks.

FIGURE 7.23

Using a paper towel to remove old grease.

3. Use phono lube to reapply the grease onto the tracks.

4. Apply a small amount of grease onto a flathead screwdriver and then apply it to the tracks, as shown in Figure 7.24.

5. After applying the grease, insert a video cassette, push Play and then Stop. Repeat this several times so that the roller guides distribute the grease evenly onto the tracks.

6. Remove any excess grease with a paper towel. You don't need to lubricate the rollers on the roller guides or the video drum.

Lubricating Worm and Cam Gears

In some models, a cam gear and/or a worm gear is located at the right top rear corner of the transport. To clean and lubricate the cam gear or worm gear, proceed to the sections in Chapter 10, "Locating lubricating points that use grease" and "Replacing old grease."

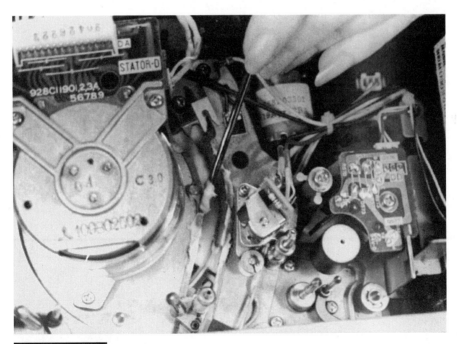

FIGURE 7.24

Applying new grease to the tracks.

FIGURE 7.25

Applying new grease to the grooves of the pinch roller cam gear.

In newer models, the lower portion of the pinch-roller assembly is a cam gear.

1. Remove the upper portion of the pinch-roller assembly.

2. To remove a pinch roller, proceed to the appropriate section in this chapter.

3. Remove the old grease from the cam gear with a paper towel and add the new grease into the grooves on the cam gear with a flathead screwdriver, as shown in Figure 7.25.

Review

To find all lubricating points:

1. Turn the power on, insert a video tape, push Play, and note all the parts that move—especially the hinges.

2. Wiggle each piece under the cassette carriage with your finger, and look for all the hinges.

3. Spin all rollers and wheels to check for freedom of movement.

4. Put a drop of household oil onto each hinge, roller, and wheel shaft using a C-, E- or O-ring.

5. Do not over lubricate; just one drop will do.

6. Do not lubricate the video drum.

7. Re-grease the tracks under the roller guides, worm gears, and cam gears.

Removing and Servicing Cassette Cartridges

A **video cassette is** the cartridge which holds the video tape. The cassette holder is where you insert the tape; it then pulls the tape into the machine. The cassette carriage is the complete assembly that cradles the cassette holder and is attached to the transport. These are the main parts covered in this chapter.

It's necessary to remove the cassette carriage to repair or replace a broken gear on the carriage, a bad tape sensor, or the housing loading motor. In some models, it's necessary to remove the carriage to reach the idler wheel and clutch assembly or to replace a tire on a wheel. It's also helpful when unjamming a jammed video cassette. Be sure to unplug the unit before proceeding. Do not plug the unit in while the carriage is out of the unit. It can cause the gears to misalign.

Preparing to Remove Cassette Carriages

To remove the cassette carriage, you first need to remove the front cover. For removing front covers, refer to the first three sections of Chapter 32. In 30% of all VCRs, a metal bracket runs across the top

FIGURE 8.1

Locating the bracket across the front of the carriage.

front of the carriage, not mounted to the carriage, but mounted to the chassis, as shown in Figure 8.1.

Most models have one screw on each end of the bracket or where the bracket completely crosses the unit, containing three screws.

1. Remove the screws and lift the bracket straight up. In a few models, the brackets might have latches holding it down (see Figure 8.1).

2. Push the latches back to release the bracket.

3. Lift the bracket straight up and off.

4. To remount the bracket, push the bracket straight down over the latches until it latches.

In other models, the screws will be mounted through the top of the front cover and through the bracket. When you've removed the front cover, the bracket will appear to have no screws. Lift the bracket straight off.

FIGURE 8.2

Latches above the bracket across the carriage.

In newer models, when you remove the front cover by releasing the latches on top of the bracket, you also release the bracket, as shown in Figure 8.2.

In older models, cassette carriages have a white, green, red, brown, or black plug, with eight to 10 wires protruding from it. This plug needs to be unplugged. Usually, the plug is located on the back right side of the carriage, on a small circuit board, as shown in Figure 8.3. In other models, the plug is on the left side at the back of the carriage. In some models, you might find two plugs on the right side, instead of one. One will be large in size and the other small. Unplug both of these plugs.

Another system used has a ribbon type wire attached to the carriage, instead of wires and a plug. To disconnect the ribbon, refer to the section in Chapter 9 "Number 5," under the subhead, "Plugs and wires to disconnect." Although the ribbon does disconnect, it isn't necessary to uncouple it. If you are not working on the carriage, but

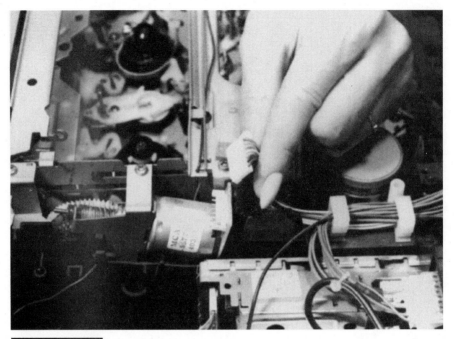

FIGURE 8.3

Locating the white plug on a small circuit board.

only removing it to get underneath the carriage, lift the carriage with the ribbon intact, flip it to the right, and lay it on top of the unit. Be sure to place a towel between the carriage and the top of the unit for insulation.

Newer models have no plugs connected to the carriage.

Cassette Carriage Removal

Six different procedures are used to remove cassette carriages in various types of VCRs. Read each of the following sections and follow the instructions that relate to the model you have.

First Type

This procedure is the most common one used to remove a cassette carriage. In older models, a plug has six to 10 wires attached to it at the back right corner of the carriage. Unplug this plug. Newer models have no plugs to unplug. At the back of the carriage, two screws are mounted to the transport, one on each side. The screws could be straight behind the carriage, as shown in Figure 8.4, or off to the side, as shown in Figure 8.5. The screws are usually gold in color, but are sometimes red or black. Remove the two screws. Slide the whole carriage back about ½ inch to release the front latches. Pick the carriage straight up and out.

Second Type

The second type of carriage is similar to the first type, except that this type has six mounting screws instead of two. Unplug the plug. Remove the four top screws mounted to the chassis, two on each side of the carriage, as shown in Figure 8.6. Remove the two screws at the back of the carriage mounted to the transport (see Figure 8.4 and Figure 8.5). Slide the carriage ½ inch back to release the latches. Lift the carriage straight up and out.

FIGURE 8.4

Locating the mounting screws behind the carriage, to the transport.

FIGURE 8.5

Another location of the mounting screws off to the side.

FIGURE 8.6

Locating the four mounting screws on top of the carriage.

Third Type

In older models, unplug the plug. Newer models have no plugs. This type of carriage is mounted with six screws. The front two screws are mounted on top of the carriage. Two screws are on each side of the carriage. Remove the outer two screws, as shown in Figure 8.7. At the back of the carriage, two screws are mounted to the transport, one on each side, (see Figures 8.4 and 8.5). Remove these two screws. On the front of the VCR are two more screws. A small circuit board might be in front of one of these screws, as shown in Figure 8.8. This little circuit board has a power stop and eject button on it. To remove this circuit board, pull back the latch at the top of the board and simultaneously pull the top of the board out. Then, lift the small circuit board up to unplug it from the mother board. Now, you can remove the two front screws and lift the carriage straight up and out.

FIGURE 8.7

Locating the four mounting screws on top of the carriage removing the outer two screws.

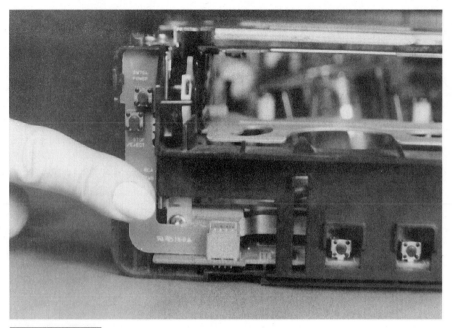

FIGURE 8.8

The mounting screw behind the circuit board.

Fourth Type

This model has a bracket on each side at the base of the carriage with three screws on each side, as shown in Figure 8.9. These brackets mount the carriage to the transport. If the unit has three large screw heads, remove only the middle screw on each side. If the unit has two small screw heads and a large screw head in the middle, then remove all three screws. Lift the carriage straight up and out, then unplug the plug on the right side of the carriage.

Fifth Type

In older models, unplug the plug. Newer models have no plugs. At the back of the carriage, two screws are mounted to the transport, one on each side, (see Figures 8.4 and 8.5). Remove these two screws. Slide the cassette carriage forward about ¼ inch. Lift the carriage straight up and out. If the carriage doesn't come out easily then you may need to remove the two screws in the front of the carriage holding the transport down, (refer to Figure 9.19). You will need to remove these

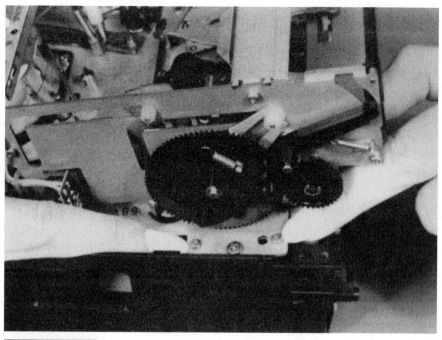

FIGURE 8.9

Locating the three mounting screws on each side of the carriage.

screws because the heads of the screws might stop the carriage from sliding forward.

Sixth Type

Unplug the plug. This type of carriage is mounted with four mounting screws. All four screws are mounted on top of the carriage and into the chassis. One screw is in each corner (see Figure 8.6). Remove the four screws and lift the carriage straight up and out.

Remounting Cassette Carriages

Older models have a loading motor with a plug on the back of the carriage (see Figure 8.3) or the carriage has a belt attached to the drive pulley from the transport, (refer to Figure 26.2).

Newer models have no loading motor mounted to the cassette carriage to receive or eject the video tape. This is done by the loading mo-

FIGURE 8.10

Locating the pins on the front of the carriage.

tor on the transport instead. You need to connect the gears on the carriage to the transport. Very important! Be sure that the cassette holder in the carriage is all the way forward. Because the power on the unit has remained off, the gears on the transport have not turned and are already aligned. Pushing the cassette holder all the way forward will align the drive gear on the carriage.

To remount the carriage:

1. Align the two pins on the bottom front portion of the carriage with the slots in the transport. Figure 8.10 is an illustration of a pin.

2. Place the pins into the slots and slide the carriage forward until the screw holes on the carriage align with the screw holes on the transport.

3. Try lifting up on the front of the carriage to be sure that both pins are latched.

4. If you didn't have to slide the carriage back to remove it, place the carriage back into its original position and align the screw holes.

5. Plug in any plugs, if needed.

6. Check the gear alignment before replacing all the screws. To check the gear alignment, read each of the following sections on gear alignment and follow the instructions that relates to the model that you have.

First Gear Alignment

This type of cassette carriage has a cassette loading arm or sliding gear at the bottom of the carriage. Align the last tooth on the sliding gear to the last slot on the drive gear mounted to the transport, as shown in Figure 8.11.

Second Gear Alignment

This type of cassette carriage has a drive gear mounted at the rear of the carriage with one larger slot and one larger tooth. On the transport is a sliding gear, also, with one larger slot and one larger tooth. Align

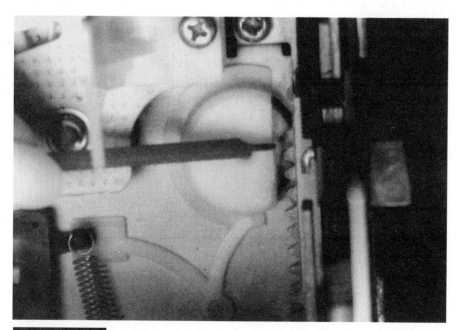

FIGURE 8.11

Aligning the last tooth on the sliding gear (right) to the last slot on the drive gear (left).

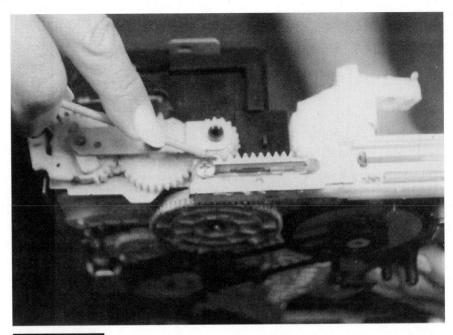

FIGURE 8.12

Aligning the large tooth on the drive gear (top) to the large tooth on the sliding gear (bottom).

the larger tooth on the sliding gear to fit into the larger slot on the drive gear and, at the same time, align the larger tooth on the drive gear to the larger slot on the sliding gear, as shown in Figure 8.12.

Third Gear Alignment

This type of cassette carriage has a drive gear protruding out the bottom of the carriage. Align the first slot of the gear on the carriage, as shown in Figure 8.13, to the last tooth on the gear mounted on the transport.

Fourth Gear Alignment

This is the most common type and aligning it is easy. On the cassette carriage, the drive gear is mounted at the bottom of the carriage, (refer to Figure 26.4). Be sure that the cassette holder is all the way forward, then place the cassette carriage back into its original position. The gears will automatically align. For more information, refer to the section in Chapter 15, "Cassette carriage alignment."

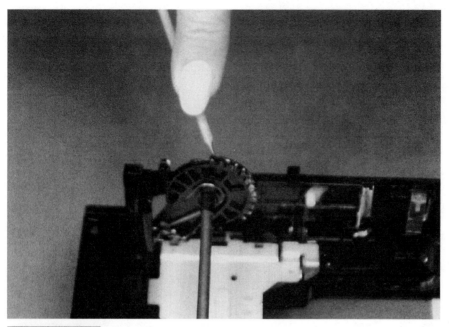

FIGURE 8.13

The first slot on the gear.

Lubricating the Cassette Carriage

In most models, you do not need to pull the cassette carriage out of the unit to lubricate it. The carriage needs to be greased, not oiled. Most all older models have a worm gear.

1. Spread the grease on the gear, as shown in Figure 8.14.

2. Insert a video cassette and push Eject to distribute the grease on the worm gear.

3. Grease the drive gears located mainly on the right side of the carriage, as shown in Figure 8.15.

The video cassette holder has four pins that protrude through the sides of the carriage. These pins run down a track when you insert a video tape. Two pins are on each side; grease each pin on each side, as shown in Figure 8.16. Then insert a video tape and push Eject to distribute the grease.

FIGURE 8.14

Spreading new grease on a worm gear.

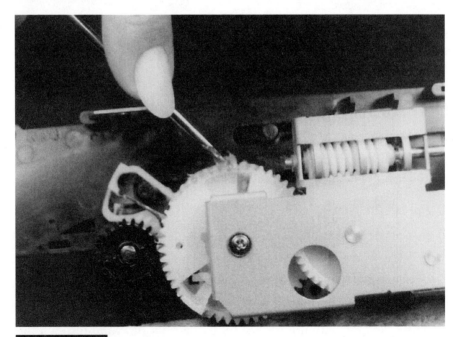

FIGURE 8.15

Spreading new grease on the teeth of the gears on the carriage.

FIGURE 8.16

Spreading new grease on the pins of the video cassette holder.

Review

1. Remove the front cover.

2. Remove the bracket across the top front of the carriage.

3. Remove the plug located at the back of the carriage.

4. Remove the mounting screws:

    A. Located at the back of the carriage, mounted to the transport.

    B. Mounted to the bracket on top, in front of the carriage.

    C. Located on top of the carriage, one in each corner, or two on each side.

    D. Brackets mounted to the base of the carriage, into the transport, on each side of the carriage.

    E. In the front bottom of the carriage, mounted to the transport. You'll need to push the cassette holder back to remove the screws.

5. In most models, push the carriage back about ½ inch to release the latches. Then remove the carriage.

6. In all other models, the carriage lifts straight out.

7. When remounting, be sure that the gear on the carriage is aligned to the gear on the transport.

8. When remounting, be sure that the pins in front of the carriage are locked in place.

9. After greasing the gears, insert a video tape to distribute the grease.

Getting Into the Undercarriage

The undercarriage contains a system of gears, motors, belts, and pulleys. To service and clean the undercarriage, remove the bottom cover plate, then open up the large circuit board. This gives you clear access to all the systems in the undercarriage.

Removing the Bottom Cover Plate

In older models, the bottom cover plate covers the entire bottom of the unit.

1. Unplug the VCR.

2. To remove the bottom cover plate, place the unit on its side.

3. Remove all the Phillips head screws on the bottom cover plate. From 3 to 12 mounting screws are on the bottom cover plate.

   Read each of the following sections and follow the instructions listed in each section that pertains to your unit.

1. In some models, once you remove the mounting screws, the plate will lift right off.

2. For most models, slide the plate straight forward about ¼ inch to clear the plastic tabs holding the plate on.

3. In a few models, slide the plate in the opposite direction, straight back about ¼ inch. Lift off the bottom cover plate.

4. In other models, remove the front cover to remove the bottom plate. The latches on the front cover go over the plate. Refer to the first three sections in Chapter 32.

5. A few models have four rubber or plastic feet, one in each corner on the bottom cover plate. In the center of each foot is a Phillips head screw. Remove the screws and each foot. This cover plate comes off. When remounting the bottom cover plate, reverse the process.

6. In newer models, the bottom cover plate covers only part of the bottom and you cannot access the undercarriage through the bottom of the unit. Here are some examples: The most common are the models with a bottom plate that covers only the very back portion of the unit. In other models, the bottom plate covers only the right side of the unit. If you have a newer model, proceed to the section in this chapter, "Transports."

Bottom Circuit Board

After removing the bottom cover plate, you'll find a large circuit board covering the undercarriage. You must open the circuit board to reach the undercarriage. All circuit boards are hinged and swing open like a door. Determine which end of the circuit board lifts out and which end is hinged.

TRADE SECRET Most circuit board hinges are located near the rear of the unit. In a few models, they're located on the side of the circuit board. Most of the circuit board mounting screws are located near the front of the unit. Some models have no circuit board covering the undercarriage. In these models, just remove the bottom cover plate.

Locating Mounting Screws

Each circuit board has five to 20 Phillips head screws. They aren't all mounting screws. The other screws serve different functions. Remove only the mounting screws to open the circuit board.

In most models, the mounting screws are at the opposite end of the circuit board from where you find the hinges. Mounting screws have special markings around the screws. The special marking is a white circle or a pointing arrow, a plastic washer under the screw, or the screws are different in color from the rest. For example, if you find metallic screws and red or blue screws, the mounting screws are the red or blue screws. If you have all gold screws, the mounting screws are the brighter more vivid color of gold. If you find only red screws, then remove them all. The most common marking used for the mounting screws is a pointing arrow or the screws are red in color.

Ways to Open a Circuit Board

Five different methods are used to open the circuit board in your VCR, after you have removed the bottom cover plate. Read each of the following sections and follow the instructions listed in each section that pertain to your unit.

First Way

Look for the hinges at the rear of the unit. In this model, the hinges are two round white plastic cylinders. The cylinders are inserted into a sliding groove, as shown in Figure 9.1. Next, locate the mounting screws. This model has three screws, with an arrow pointing at each one, as shown in Figure 9.2. Remove only these three screws. At the bottom of the front cover, open the small panel doors, as shown in Figure 9.3. (I've removed the panel door to show the controls protruding through the front cover above the circuit board.)

If the controls are within ½ inch of the circuit board, the controls are attached. If any controls are near the circuit board, pull the board straight back approximately ½ inch, so the controls can clear the front cover, as shown in Figure 9.4. Now, open the circuit board like a door, as shown in Figure 9.5.

> **QUICK»TIP**
>
> If the circuit board doesn't slide back and open easily, you have a different type of model than is shown. Another mounting screw might need to be removed.

Second Way

This model has two shafts made of black plastic coming in from the chassis, one

Locating the plastic cylinder hinges.

An arrow pointing at each mounting screw.

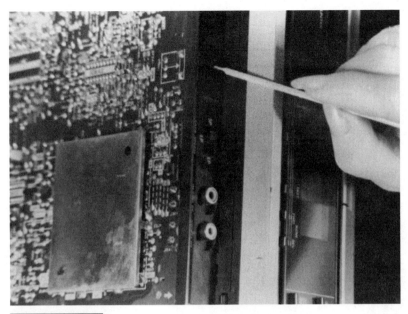

FIGURE 9.3

The controls protruding through the front cover.

FIGURE 9.4

The controls clearing the front cover.

FIGURE 9.5

Opening the circuit board like a door.

on each side of the circuit board. The hinge is a long metal bracket along the back of the circuit board. Each end of the bracket is inserted around the black plastic shaft, as shown in Figure 9.6. Each mounting screw has a white circle painted around it, as shown in Figure 9.7. Remove these screws. Check the front cover for a small panel door, as described in the previous section. If it has no panel door, lift up on the front of the circuit board.

Third Way

This model has two to four white plastic hinges across the back of the circuit board, as shown in Figure 9.8. The mounting screws are red in color and are near the front of the unit. Remove only the red screws. After removing the screws, you'll see four to six black plastic clips around the outside of the circuit board (refer to Figure 3.12). You

FIGURE 9.6

A black plastic shaft and a metal bracket hinge.

FIGURE 9.7

A mounting screw with a circle around it.

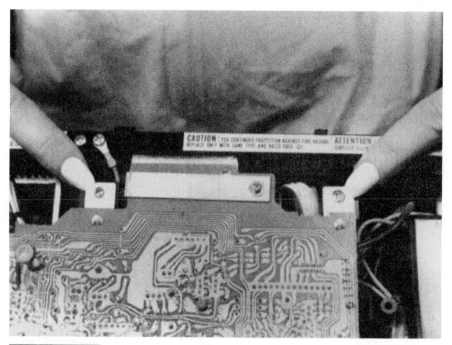

FIGURE 9.8

The white plastic hinges.

might find two black plastic clips on the outside and one or two in the middle of the circuit board, protruding through a small rectangular hole. To release the clips, pull back on the clip and simultaneously pull up on that portion of the circuit board. Release each clip until the circuit board opens.

Fourth Way

This model has two large circuit boards beside each other. They are held together with brackets. Approximately 19 screws are on these circuit boards. The screws are usually gold or black in color. One screw is red or blue and is mounted through the middle of the front bracket. Remove this different-colored screw, as shown in Figure 9.9. Next, look for any small panel doors on the front cover. Refer to Figure 9.3 to visualize the controls protruding out the front cover. Pull the circuit board back to clear the controls and open the circuit board.

FIGURE 9.9

The red or blue mounting screw.

In a few models, a small, black plastic clip sticks out through a metal bracket along the front of the circuit boards. The clip is in front of the screw that you just removed. Push the clip down to release the circuit board. Then, pull the board straight back to clear the controls. The circuit board will open like a door.

Fifth Way

If no small panel door is on the bottom of the front cover, you've removed the mounting screws, and it still won't open, the front circuit board is attached to the bottom circuit board. First, remove the front cover. Proceed to the first three sections of Chapter 32. After removing the front cover, you'll find four to six black plastic clips across the top edge of the front circuit board, as shown in Figure 9.10. Push up on each clip one at a time, simultaneously pulling forward on the circuit board near that clip. After releasing all the clips, pull the bottom of the front circuit board toward you about ½ inch. Moving the board will clear any tabs holding the front circuit board to the chassis. In this

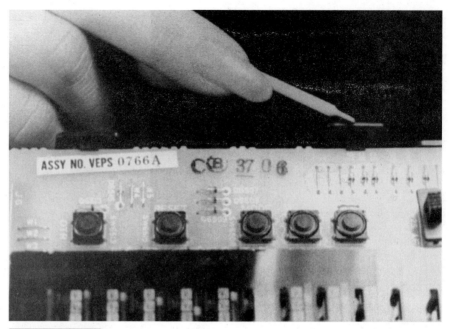

ASSY NO. VEPS 0766A   CB 37 06

FIGURE 9.10

The black plastic clip.

model, the front and bottom circuit boards open together like a door, as shown in Figure 9.11.

If the Circuit Board Still Won't Open

If, after removing the mounting screws, the circuit board still won't open, look on the circuit board for black plastic clips that hold it down. If you start to open the circuit board and it bends in the middle, check each hinge for a locking red screw in front of the hinge. Figure 9.6 shows the type of bracket where you would find these red screws. Just remove the locking screws.

Check two more places if the circuit board still won't open. Either you have missed a mounting screw or someone has put a wrong-colored screw in the wrong screw hole. If the screws have been switched, remove all of the screws. Try to open the circuit board every time you take out a screw. It's important to remember which screw came from which hole. Each screw has a different thread and you don't want to strip the threads. (Refer to the section in Chapter 1, "Information on screws.")

FIGURE 9.11

The front, and bottom circuit boards opening together.

Transports

In newer models, to service or repair the undercarriage, you need to remove the transport through the top of the unit. In most models, the cassette carriage can stay attached to the transport while removing the transport. Read each of the following sections and follow the instructions listed in each section that pertains to your unit.

Plugs and Wires to Disconnect

Before removing the transport, you must unplug any plug that is attached to the mother board. It could have anywhere from zero to four plugs. Read the following sections to find where the plugs are located.

1. A grounding wire might be attached to a screw located at the left rear corner of the transport, as shown in Figure 9.12. Remove this screw and slide the grounding wire over to the side.
2. A plug on top of the eraser head and the wires from the plug attach to the mother board, as shown in Figure 9.13. Simply unplug the plug on top of the eraser head.

FIGURE 9.12

Grounding wire mounted to transport.

FIGURE 9.13

Plug on top of eraser head.

3. A drum motor plug might be at the rear of the video drum, as shown in Figure 9.14. Unplug this plug.

4. In most models, a ribbon protrudes out from the back of the lower video drum. This ribbon is plugged into the video head amplifier. If the amplifier is or is not attached to the transport unplug the ribbon by pulling it out of the plug, (see Figure 9.15). On the other hand, if the amplifier is mounted on top of the transport, again you need to unplug the ribbon going to the drum. On top of the amplifier, two ribbons protrude out and are attached to the mother board. Remove the amplifier and drape the amplifier over the back of the unit, as shown in Figure 9.16.

5. A plug might be on top of the A/C head and a ribbon-type wire might be attached to the plug going to the mother board, as shown in Figure 9.17. To unplug the ribbon, lift up on the top latch of the plug and pull the ribbon out. To reconnect the ribbon, push the bare wires on the end of the ribbon into the plug and push the latch down to lock the ribbon into position.

FIGURE 9.14

Drum motor plug at rear of video drum.

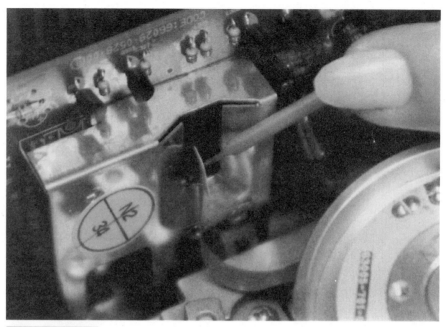

FIGURE 9.15

The ribbon from the video drum is plugged into the video head amplifier.

FIGURE 9.16

Draping the video head amplifier and ribbons over the back of the unit.

FIGURE 9.17

Releasing the top latch on the pins above the A/C head.

Mounting Screws

The following tells you where all the mounting screws are located on a transport. Each transport has a different pattern of mounting screws. Read all nine sections so that you can figure out the mounting pattern of your VCR. Then, remove all the mounting screws.

1. A mounting screw might be on each side, on top of the cassette carriage. Remove these two screws. Or, two screws could be on each side. Remove the outer screw (refer to Figure 8.7).

2. Look at the front of the transport under the carriage. A hole might run through the middle of the cassette holder. Under the hole will be a mounting screw on the transport, as shown in Figure 9.18.

3. Two mounting screws with arrows pointing at them might be in front of the cassette carriage, as shown in Figure 9.19.

4. Two mounting screws might be in front, one in each corner under the cassette carriage, but you will need to pull the carriage out before you can remove these screws.

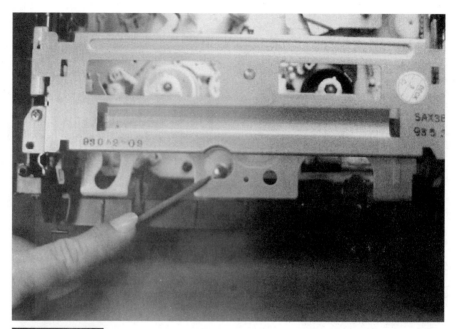

FIGURE 9.18

The hole through the cassette holder showing **mounting screws.**

FIGURE 9.19

The mounting screws at the front of the carriage with arrows pointing at them.

5. Four screws, two on each side of the transport, might be mounted to the chassis, as shown in Figure 9.20. (I drew arrows pointing at the mounting screws.)

6. Two screws might be mounted on top of the transport, near the back, one in each corner, as shown in Figure 9.21.

7. One screw might be in the middle on a bracket behind the transport, as shown in Figure 9.22.

8. In some models, a video head amplifier is mounted to the transport and two mounting screws are at the base of the amplifier near each end. In other models, another mounting screw might be in the middle, behind the amplifier, near the top, as shown in Figure 9.23. (I drew arrows pointing at the three mounting screws.)

9. In a few models, two mounting screws protrude up through the mother board with white circles around them, or arrows pointing at them as shown in Figure 9.24. You need to remove the bottom plate cover. In other models, three mounting screws protrude up through the bottom of the unit and the mother board with no cover plate.

FIGURE 9.20

The mounting screws on the sides of the transport.

FIGURE 9.21

A mounting screw on top and near the back of the transport.

FIGURE 9.22

A mounting screw on a bracket behind the transport.

FIGURE 9.23

Three mounting screws to the video head amplifier.

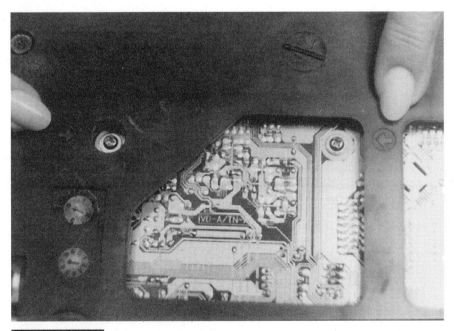

FIGURE 9.24

Mounting screws under the mother board with arrows or white circles around them.

FIGURE 9.25

Transport plugs under the transport.

Removing the Transport

To remove the transport:

1. Take hold of the front of the transport and pick up on it to see if it feels loose. If not, look for more mounting screws. Do not try to remove the transport at this time. In most models, the front of the transport will come up easily. At the back of the transport, plugs are underneath. These plugs are attached to the bottom of the transport and are plugged into the mother board, as shown in Figure 9.25. If a video head amplifier is attached to the transport or on top of the transport, you need to remove the amplifier first before removing the transport.

2. Check and be sure that you have removed all mounting screws.

3. If the video amplifier is attached to the transport, but not on top (refer to Figure 9.23), you need to pull hard on the amplifier to release the plugs from the mother board.

4. Place the amplifier off to the side.

5. If the amplifier is mounted on top of the transport, a screw is at the base on each end of the amplifier. Remove these two screws.

6. Look on top of the amplifier for two ribbons protruding out, attached to the mother board.

7. Lift the amplifier and drape the amplifier over the back of the unit (see Figure 9.16).

> **QUICK»TIP**
>
> Remember, keep the transport level when removing it. In some models, the LED and tape sensors are mounted to the mother board, (refer to Figure 21.1 and Figure 21.7). They protrude up through the transport. Be careful not to damage these sensors when removing the transport.

8. When removing the transport, hold it near the back on each side. Keep the transport level so that you will not bend any pins in the plugs under the transport. Sometimes you need to pull up hard to release the plugs, but not too hard. Figure 9.26 shows the location of some of the plugs on the mother board under the transport. (I drew arrows pointing at the plugs.) Plugs can be located almost anywhere under the transport.

9. A few models have latches at the back of the transport, as shown in Figure 9.27. To remove this transport, lift up on the front of the transport and the latches at the rear will disconnect by themselves and the transport will lift out.

FIGURE 9.26

The plugs on the mother board for the transport.

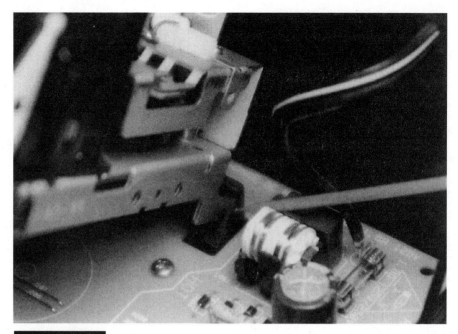

FIGURE 9.27

Latch at back of the transport.

Remounting the Transport

To remount the transport:

1. Place the transport over the mother board and align (if any) the LED or tape sensors mounted to the mother board. Refer to Figure 21.7.

2. Slide the sensors up through the holes in the transport. Refer to Figure 21.1.

3. Four or five alignment pins are above the mother board that the transport sits on, as shown in Figure 9.28. (I drew arrows pointing at the pins.) Align these pins to the holes on the transport, which are beside the mounting screw holes. The plugs under the transport will automatically align when you align the pins to the pin holes.

4. Push the transport straight down, specifically pushing at the rear of the transport until you feel the plugs snap back together or the mounting holes on the transport are up against the chassis mounting screw holes.

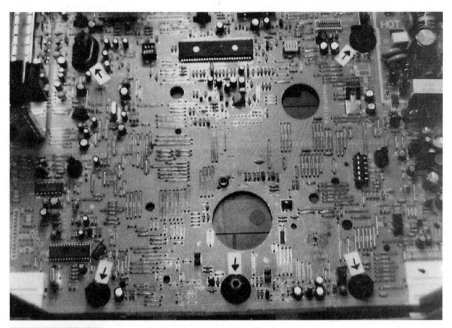

FIGURE 9.28

Alignment pins above mother board for the transport.

If you have a model with latches at the rear of the transport, simply:

1. Lower the back of the transport and slide the tab on the transport into the latch.

2. Lower the front on the transport until the front of the transport lays flat on the mother board.

3. After remounting the transport, remount the video head amplifier.

4. If the video amplifier plugs into the mother board, align the pins on the board to the pin holes on the bottom of the video amplifier and push down until the plugs snap into place (refer to Figure 31.3).

5. Connect the ribbon going to the video drum. If the video amplifier is mounted on top of the transport, place it back into its original position and replace all mounting screws.

6. Replace all mounting screws to the transport and plug in all plugs.

Review

To get into the undercarriage

1. Remove the bottom cover plate screw, then slide the plate forward for removal.

2. To open a circuit board:
   A. Look for the hinges located at the rear of the unit.
   B. Look for the mounting screws, which are usually located at the opposite end from the hinges.
   C. The mounting screws are marked by:
      1. a white circle around them
      2. an arrow pointing at them
      3. a plastic washer under them
      4. a different color from the rest

3. Look for a small panel door located at the bottom of the front cover. If there is a door, pull the circuit board back before you pull it open.

4. After removing all mounting screws, be sure to release any black plastic clips.

5. Always check the bracket across the front of the circuit board for any clips that are keeping the board from opening.

6. If the circuit board has no panel door and doesn't open after removing all the mounting screws, the front circuit board comes open with the bottom circuit board.

7. Check the hinges for locking red screws located directly in front of the hinge.

To remove transports

1. When removing a transport, be sure to unplug all the plugs that are attached to the top of the transport.

2. After removing all of the mounting screws, lift up on the front of the transport to see if the transport is loose.

3. Remove the transport by holding both sides at the rear of the transport, pulling it up, and keeping the transport level.

4. When remounting the transport, slide any tape sensors through the holes in the transport and align up the chassis pins to the pin holes in the transport.

5. Do not push down on the transport until everything looks aligned. Then, push harder on the rear of the transport until you feel the plugs snap back together under the transport.

6. Be sure to plug in all of the plugs on top of the transport.

Servicing & Lubricating the Undercarriage

This chapter completely lists all the moving parts in the undercarriage. Servicing and lubricating all of the moving parts are very important for proper maintenance in a VCR.

Please note, as you read further in this chapter, and before removing the transport, that you must insert a blank video tape and leave it in the unit. This is important because, when re-lubricating cam gears, you will need to rotate the gear; this cannot be done unless the cassette holder is in the "down" position. To remove the transport, refer to the section in Chapter 9, "Transports."

Belts in VCRs

In older models, the transport uses belts to drive most of the moving parts. Refer to the section in Chapter 9, "Getting into the undercarriage." Locate the main drive pulley mounted to the shaft on the capstan motor. The capstan motor drives the capstan shaft, Fast Forward, Rewind, loads the videotape, and, in some models, drives the cassette carriage. You'll find anywhere from two to four belts coming off this

FIGURE 10.1

The main drive pulley.

Note the belt configurations and remember their proper positions before you remove them. Each model has a different configuration, but the basic idea is exactly the same.

pulley, as shown in Figure 10.1. Also, a pulley is mounted on the drum motor and a belt drives the upper video drum. In these older models, the belts slip off of their pulleys. No brackets are in the way.

Newer models have only one belt, which is mounted to the pulley on the flywheel and the flywheel is at the base of the capstan motor, as shown in Figure 10.2. This belt drives the idler wheel and clutch assembly (Fast Forward, Rewind, and take-up). In some models, to access this belt, you need to remove the transport. Any belts that are blocked by a mounting bracket, or you can't remove, refer to the section in Chapter 19, "Removing loading motors and belts" and to the section in Chapter 25, "Removing the capstan shaft in belt-driven VCRs."

Cleaning Pulleys

Saturate a glass brush in either cleaning alcohol or head cleaner (refer to the section in Chapter 1, "Making a glass brush").

FIGURE 10.2

Pulley on the fly-wheel and drive belt for the clutch assembly.

1. Place your finger on the opposite side of the pulley to hold it in place while cleaning.
2. Place the glass brush in the groove of the pulley.
3. Rub back and forth with a scrubbing motion, as shown in Figure 10.3.
4. Clean the brush by using a paper towel, rotate the pulley, and clean the next section. Continue this procedure until you've cleaned all the way around the pulley.

Clean each pulley in the same manner.

Checking and Cleaning Belts

1. Fold a clean paper towel in half. Saturate one corner of the towel in the cleaning solution.
2. Place the belt on the corner of the saturated towel and fold the towel over the belt tightly, as shown in Figure 10.4.
3. Pull the entire length of the belt through the paper towel in short intervals. You'll notice a lot of black residue coming off of the belt. Periodically, reposition the belt in the towel.

FIGURE 10.3

Cleaning a pulley with a glass brush.

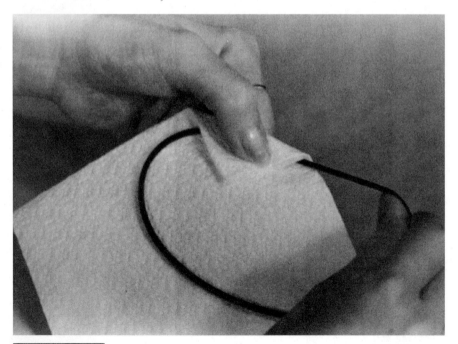

FIGURE 10.4

Cleaning a belt.

Continue this procedure until there's no black residue coming off the belt. Clean all the belts in the same manner.

Checking and Lubricating Pulleys

Before replacing the belts onto their pulleys, check the pulleys for lubrication. If the pulley is attached to a worm gear, as shown in Figure 10.5, place a drop of oil onto its shaft between the pulley and the worm gear. For all other pulleys, spin them to be sure that they are moving freely. If a pulley doesn't move freely, the shaft to the pulley needs to be lubricated.

On a double pulley:

1. Where two belts are attached, remove the E- or O-ring (refer to the section in Chapter 7, "Removing E- or O-rings"), lift the pulley off, and clean the shaft with a paper towel.

FIGURE 10.5

Pulley and belt attached to a worm gear.

2. Place a drop of oil onto the shaft and remount the pulley.

3. If the pulley is attached to a clutch assembly, you need to remove the clutch assembly (refer to the section in Chapter 23, "Removing clutch assemblies").

4. Clean the shaft and place a drop of oil onto the shaft.

5. Remount the clutch. If the shaft is solidly mounted to the pulley and it is hard to turn, it has a build-up of residue between the shaft and the inside of the assembly around the shaft. On the other hand, it could have lost its lubrication and become dry.

6. Remedy this problem by spraying degreaser into the shaft at the point shown in Figure 10.6 (I've removed a pulley and clutch assembly for demonstration purposes only).

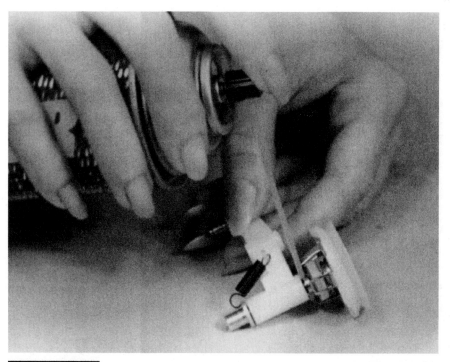

FIGURE 10.6

The point to spray degreaser.

7. After spraying the shaft, spin the pulley and continue this process until the pulley turns freely.

8. Lubricate the shaft. Take a small flathead screwdriver and place a drop of household oil onto its tip.

9. Place the dry side of the tip onto the shaft at the base of the pulley, at the point where the shaft goes through the assembly arm, as shown in Figure 10.7.

10. Tip the screwdriver up, allowing the oil to drop down onto the shaft.

11. Go to the opposite end of the assembly, where a wheel is attached to the shaft and lubricate the point between the wheel and the assembly arm. Be careful not to get any oil on the small wheel.

12. Spin the pulley to work the oil up inside the arm onto its shaft. Repeat this procedure until the pulley spins freely.

13. Now, begin replacing the belts onto their proper pulleys. Be sure that each belt fits securely and snugly. If the belt is loose or droops, then it is stretched out and has to be replaced.

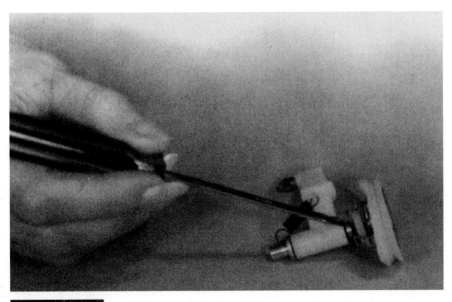

FIGURE 10.7

Placing a drop of oil onto the shaft.

In older models, after replacing all the belts onto the pulleys, leave the VCR on its side with the undercarriage exposed. Plug the VCR in and turn it on. Insert the videotape, push Play, then Stop, push Fast Forward, then Stop, push Rewind, and then Stop. Each time you push a button, check the belts that are turning to see if any of the belts are slipping. If so, replace them.

In newer models, you will need to remove the transport to get into the undercarriage. Only one belt is under the transport and it's attached to the clutch assembly (refer to Figure 10.2). After cleaning the belt and pulley, place your finger on the outside of the flywheel and turn it. Using your other hand, place your finger onto the clutch pulley and stop it from turning. If the belt is good, the clutch pulley will be hard to stop. If it stops easily, it will be stretched out and will slip on one of the pulleys. You need to replace it. To replace the belts, refer to the section in Chapter 1, "Parts."

Locating Lubricating Points that Use Oil

In older models:

1. Leave the VCR on its side so that you can observe all the moving parts mounted to the undercarriage.

2. Plug in the unit, insert the videotape, and push Play. Keep an eye on the undercarriage. As the videotape loads onto its tape path, you'll be able to see how all the moving parts operate.

3. Stop and start the unit two or three times to find all the lubricating points.

4. Remove the videocassette, unplug the unit, and lay the unit upside down.

In newer models, you need to remove the transport to get to the undercarriage. The moving parts in these sections are interconnected, usually by a hinge. A C-, E-, or O-ring is at each hinge. Lubricate them by placing a drop of household oil on top and in the split of the C- or E-ring, as shown in Figure 10.8. To lubricate under an O-ring, place a small flathead screwdriver under one side of the split in the O-ring and pry that side up (refer to Figure 7-4). Get a drop of oil under it, and then slide the O-ring back down over the shaft. Don't get any oil on the belts or pulleys that you just cleaned.

A lubricating point.

To lubricate the gears, look for a C-, E-, or O-ring attached to the shaft in the center of the gear and place one drop of oil behind the ring onto the shaft, as shown in Figure 10.9.

Direct-drive VCRs

> **QUICK»TIP**
>
> Don't remove these gears because they are perfectly aligned to each other. Any alteration of the gears will cause the unit to jam. Don't try to spin or turn these gears as you did the pulleys.

Direct-drive systems have no belts, pulleys, or gears to be cleaned or lubricated, as shown in Figure 10.10. Direct-drive systems require no servicing to the undercarriage.

Locating Lubricating Points that Use Grease

After reading this section, you should refer to the section in this chapter, "Replacing old grease."

If you have an older unit, place the VCR back on its side, plug it in, insert a blank cartridge or videotape, and push play. As the unit loads, observe all the moving parts. In newer units, the same types of parts are located under the transport and you must remove the transport to

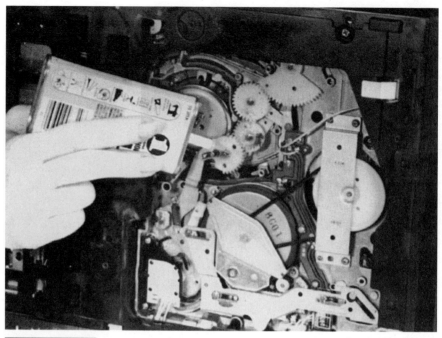

FIGURE 10.9

Lubricating the shaft of a gear.

QUICK>>>TIP If you find a trail of old black grease, you need to re-lubricate.

service these parts. You're looking for any brackets or levers that slide back and forth. Look for any moving brackets that make slight contact against the transport. Look for any rivets that protrude through a slot in a bracket. The bracket moves back and forth, leaving a trail of old grease, as shown in Figure 10.11.

Check the grooves on all cam gears. They can be located behind a small lever that moves with the cam gear, as shown in Figure 10.12. Other cam gears can be located under a mounting bracket and you can only see part of the cam gear, as shown in Figure 10.13. If the cam gear has black grease or very little grease in the grooves of the gear, you need to replace the grease.

The last place to check is the worm gears. Some worm gears are directly connected to the loading motor and other worm gears are driven by a belt and pulley (see Figure 10.5).

After a period of time, the grease can fill up with dust and become gooey. The grease also can wear thin or become hard, causing the VCR to overwork. The effect is a slipping or broken drive belt or broken gears.

FIGURE 10.10

A direct drive system.

Replacing Old Grease

Three different methods are used to replace grease. Read all of the following sections and follow the appropriate instructions.

SLIDING LEVERS AND BRACKETS

To remove the old black grease from a sliding bracket (see Figure 10.11):

1. Lay the unit or transport on its side, and, with a can of degreaser, spray off the old grease.

2. Wipe up the residue with a paper towel.

3. The next step is to take some phono lube and place a small amount on the tip of a small flathead screwdriver. Spread the grease onto any moving bracket from which you removed the old grease.

FIGURE 10.11

A lubricating point that uses grease.

CAM GEARS

To replace the grease in a cam gear, place a small flathead screwdriver inside a paper towel and remove the old grease in the grooves, as shown in Figure 10.14. Another approach is to use a can of degreaser. Spray each groove until it's free of the old grease and use a paper towel to absorb the residue. Next, spread some phono lube into the grooves of the cam gear, as shown in Figure 10.15.

In units with a mounting bracket or sliding levers over the top of the cam gear, you can only clean and grease the part of the gear that is

FIGURE 10.12

The grooves in a cam gear.

FIGURE 10.13

Cam gear protruding out from behind a mounting bracket.

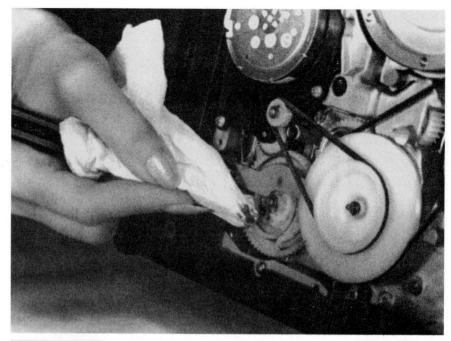

FIGURE 10.14

Removing old grease from the grooves in a cam gear.

FIGURE 10.15

Spreading new grease in the grooves of a cam gear.

sticking out (see Figure 10.13). To remedy this problem, read the next two following sections:

1. In newer models, insert a blank cartridge into the unit and unplug it. The cassette holder has to be in the down position to rotate the cam gear. Then remove the transport. To rotate the cam gear, refer to the sections in Chapter 19, "Loading motors" and "Loading motors and belts." Then, place your finger on the drive pulley or worm gear attached to the loading motor and rotate it to reposition the cam gear. Now you can clean and grease this section of the cam gear. Continue rotating the gear until you go all the way around.

2. In older models, which you can get into the undercarriage, plug it in, insert a blank cartridge or videotape, and push Play to make the gear move. When the newly cleaned section of the cam gear goes under the bracket, unplug the VCR and it will stop. Clean this section of the cam gear that is sticking out. Plug the unit back in and it will unload and stop. Push Play again and wait until the VCR loads all the way and stops. Now you can clean the last part of the cam gear. To apply the grease, push Play; while the gear is moving, place grease onto the groove as it passes the open area. Push Stop to move the gear in the opposite direction. You might need to repeat this process a couple of times to distribute the grease evenly.

WORM GEARS

To clean the worm gear:

1. Spray it with degreaser and wipe the residue off with a paper towel.

2. To lubricate the worm gear, spread the grease on the gear (refer to Figure 8.14).
   A. If the worm gear is on the cassette carriage, insert a videocassette and push Eject to distribute the grease on the worm gear.
   B. If the worm gear is in the undercarriage, push Play and Stop to distribute the grease.
   C. If you had to remove the transport, then rotate the worm gear by placing your finger on the drive pulley or worm gear attached to the loading motor.

Review

1. Check for any broken belts.

2. Examine and remember the position of each belt before removing it.

3. Be sure to clean all belts and pulleys.

4. While cleaning the belts, give them the stretch test.

5. Spin all the pulleys to make sure they spin freely.

6. Lubricate all the pulley shafts.

7. Replace the belts back onto their proper pulleys.

8. Lubricate all gear shafts.

9. Do not over lubricate. One drop goes a long way.

10. Check the grease points for lubrication.

11. Lubricate all moving brackets and levers.

12. Lubricate all cam gears and worm gears.

Setting the Torque on a Video Tape

The back tension guide keeps the correct amount of torque on the videotape while it is playing. Torque is the amount of pressure being applied. The back-tension guide is located on the videotape path directly in front of the supply tape guide, as shown in Figure 11.1. It's mounted on a moving bracket.

Detecting Back-Tension Guide Problems

You have to determine whether the problem is with the back-tension guide, video head, or A/C head. These three problems have some common symptoms. To find out, connect the TV monitor to the VCR and insert a rented videotape (refer to the

The wrong amount of torque occurs if:

1. **The bracket can become bent or hung up causing no torque on the videotape.**

2. **The spring attached to the tension guide bracket can become stretched out after a period of time causing too little torque on the videotape.**

3. **The felt on the brake shoe band wears down or comes off, causing none or too little torque on the videotape.**

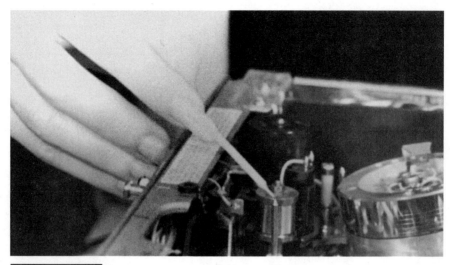

FIGURE 11.1

The back tension guide pole.

section in Chapter 5, "Connecting the VCR to your TV monitor"). Push Play and watch and listen to the TV monitor. If the video head is dirty, the picture will be partially distorted or completely snowy. Refer to the section in Chapter 5, "Dirty video heads." If the A/C head is out of alignment, a belt of lines will start at the bottom of the picture and move up the screen. Proceed to the section in Chapter 20, "Lines floating through the picture" under the subhead "Video."

Checking the Tension Guide Pole

While the unit is in play, see if the center of the tension guide pole is approximately 1 to 5 mm left of the center of the supply tape guide. The supply tape guide is directly behind the guide pole.

The tape guide pole should be perfectly straight up and down, not tilted. If the pole is not tilted, place your finger on top of it while the unit is running, as shown in Figure 11.2. Move the tension guide pole approximately ½ inch to the left, as you face the front of the unit. Check for freedom of movement by releasing the tension guide pole to see if it pops back to its normal position. Now push the tension guide pole to the right for an inch or so to check for freedom of movement in this direction. Again, release the tension guide pole to see if it pops back into its normal position. If it's not bent, the tension guide pole will pop back. By following this procedure, you'll learn whether the tension guide bracket is bent, hung up, or tilted.

If the tension bracket is bent or tilted, then you need to remove the bracket and straighten it out. Proceed to the section in this chapter, "Removing a tension guide bracket." On the other hand, if you find that the tension guide pole isn't hung up or tilted, follow the steps in the following section.

Using a TV Monitor to Check the Tension Guide Pole

1. Be sure that the TV monitor is connected properly to the VCR and push Play.

2. Place your finger onto the tension guide pole as previously shown.

3. Start by slowly pushing the guide pole to the left, periodically stopping approximately every ⅛ inch and holding that position for 20 to 30 seconds. Each time you stop, check to see if the picture clears up on the TV monitor. Continue this procedure until you can go no further.

4. If the picture doesn't clear up, release the guide pole and let it pop back to its normal position. Then slowly push it to the right, again periodically stopping to check the picture. Usually, a spot to the left of the normal position will clear up the picture.

When using this method, remember that the picture must completely clear up,

QUICK REPAIR

There are three ways to detect if the tension guide is hung up or out of adjustment:

1. The TV monitor shows a completely washed out, distorted picture, or many lines and dashes are across the picture. The audio portion has a buzz or distortion in it.

2. The picture is stable with a normal clear picture and then becomes unstable with a distorted picture, but returns to a clear picture only to distort again. Each cycle takes between 5 and 15 seconds. The audio portion also will sound like it's speeding up then slowing down, or the audio might have a distortion in it that clears up along with the picture.

3. It shows a configuration of white rectangular and square boxes that resemble a large city. The configuration is a long and narrow bar across the screen that starts at the top of the screen, works its way down, then disappears. It keeps repeating this process. It might also remain stationary at the top of the screen, appearing and disappearing.

TRADE SECRET

Older models have a tension roller instead of a tape guide (Figure 11.1). The tension roller varies in width and looks like an empty spool of thread. In newer models, the supply tape guides are post mounted to the transport, one in front and one behind the guide pole, (refer to Figure 11.3).

FIGURE 11.2

Placing a finger on top of the tension guide pole.

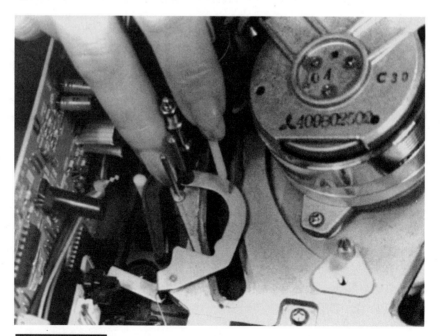

FIGURE 11.3

The point the guide pole bracket gets bent or twisted.

or stop pulsating, while holding the tension guide pole in one of these positions. If it does, proceed to the section in this chapter, "Back tension guide adjustment." On the other hand, if you can't stop the pulsating, refer to the section in Chapter 20, "Lines floating through the picture" under the subhead "Video." If you can't clear up the whole picture and stationary lines are running horizontally at the top or bottom of the picture, then your roller guides are probably out of alignment. Do not make this adjustment. Do a video tape path alignment instead. Proceed to Chapter 12, "Video tape path alignment."

Back-tension Guide Adjustment

The majority of all models have a spring connected to the tension guide-pole bracket. If you had to move the tension guide pole to the left to clear the picture, then you'll need to tighten the spring. On the other hand, if you had to move the guide pole to the right, then you'll need to loosen the spring. You'll find the adjustment at the base of the spring. The torque on the videotape is important. Just make little adjustments, one at a time, until the picture clears up. Do not overtighten the spring. Read the following types of adjustments and find the one pertaining to your unit.

First Type

In some models, the adjustment is a white plastic bracket that slides back and forth. It has notches on it and a screw in the middle, as shown in Figure 11.4. (I've removed the cassette carriage for a better view. It's not necessary to remove the carriage to make these adjustments.)

In other models, the adjustment is a metal bracket that also slides back and forth. It has notches on it and a screw in the middle, as shown in Figure 11.5.

To make either of these adjustments:

1. Eject the videocassette and loosen the screw on the bracket.
2. Move the bracket one or two notches toward or away from the spring. Usually, you'll be tightening the spring.
3. Retighten the screw, insert a videocassette, and press Play.
4. Now look at the TV monitor to see if you have a clear picture.
5. If not, re-adjust the spring tighter, a notch or two at a time.

FIGURE 11.4

First type of spring adjustment.

FIGURE 11.5

Another type of spring adjustment.

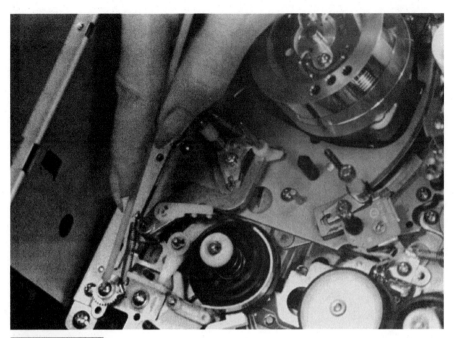

FIGURE 11.6

Second type of spring adjustment.

Second Type

In a few models, the adjustment turns back and forth. It has notches on it and a screw in the middle, as shown in Figure 11.6. In these models, the adjustment can be made while the unit is in Play.

1. Loosen the holding screw and push Play.

2. If you moved the tension guide pole to the left, move the bracket away from the spring a notch at a time.

3. Stop for a minute or so to see if the picture stabilizes or clears up.

4. If not, move it another notch and stop, repeating this procedure until the picture clears up. Do not overtighten the spring.

Third Type

In most models, the spring is attached directly to an arm mounted to the transport with notches in it, as shown in Figure 11.7. Use the same procedure to adjust the spring, but this time, move the spring itself

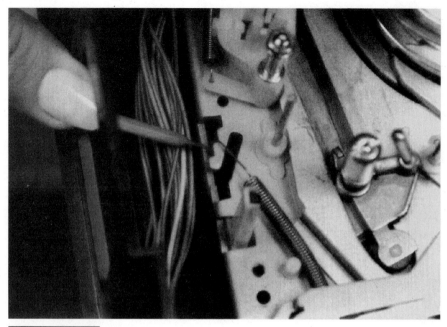

FIGURE 11.7

Third type of spring adjustment.

In some older models, you can clear the picture by moving the tension guide pole over. When you make the adjustment, the spring is tightened all the way. In this case, remove the spring and take it to a local electronics supply store for replacement or refer to the section in Chapter 1, "Parts." After replacing the spring, set the adjustment all the way toward the spring. By doing this, you are relaxing the spring. Now, follow the previous procedure and tighten the spring notch by notch until the picture clears up.

from notch to notch. Remember to check the picture each time that you move the spring until the picture clears up.

Fourth Type

Newer models have a tension spring, but there is no adjustment. In some models, there is no tension spring. In either case, you need to adjust the brake shoe band to correct the problem. Proceed to the section in this chapter entitled "Brake shoe band adjustment."

Brake Shoe Band Adjustment

When the unit is in the Play mode, the brake shoe band puts pressure on the supply spindle and causes drag. The drag causes torque on the

FIGURE 11.8

Brake shoe band.

videotape. The brake shoe band is wrapped around the supply spindle with felt inside it, as shown in Figure 11.8. The majority of the
time, you don't have to make this adjustment. If you have tried to
make the proper back-tension guide adjustment or there is no back-
tension adjustment and the only way to clear the picture is by applying pressure to the guide pole, then you'll need to do the brake shoe
band adjustment. A *Tentelo Meter* is helpful, but it's quite an invest-
ment. This tool measures the grams of pressure on videotape. You can
set the torque without this meter. When making the brake shoe band
adjustment, only move the adjustment ¹⁄₃₂ inch at a time. This adjustment is very crucial. If you overtighten this adjustment, the video
tape will not move in the play mode, like being in pause, or the tape
will move very slow and cause the tape to crinkle at the take-up tape
guide and eaten at the pinch roller.

In all units that have a tension-spring adjustment, first move that
adjustment to its center position. Read the following sections and follow the instructions that pertain to the model you have.

FIGURE 11.9

First type of brake shoe band adjustment.

First Type

In some models, the adjustment is a plastic or metal bracket with several little notches and a holding screw, as shown in Figure 11.9. Loosen the brake shoe band screw, move the adjustment one notch to the left, and re-tighten the screw. In other models, they have the same kind of bracket, but there are no notches.

In this case:

1. Move the bracket about $\frac{1}{32}$ inch to the left and re-tighten the screw.

2. Insert a videocassette and push Play.

3. See if the picture has cleared up. If not, then move the band adjustment one more notch or $\frac{1}{32}$ inch over to the left and check the picture again.

Second Type

In other models, the adjustment is a slot in the brake shoe bracket and another slot in a plastic bracket mounted to the transport, as shown in Figure 11.10.

In this case:

1. Loosen the brake shoe band locking screw, take a small flathead screwdriver and place it into the slot. Turn the screwdriver coun-

terclockwise to tighten the band. Turn the adjustment ⅟₃₂ inch and re-tighten the screw.

2. Insert a videocassette and push Play.

3. See if the picture has cleared up. If not, then move the band adjustment another ⅟₃₂ inch and check the picture again.

Third Type

In a few models, the brake shoe band is connected to a tab mounted to the transport; the tab is the adjustment.

1. Use needle-nose pliers and place the pliers over the tip of the tab, as shown in Figure 11.11.

2. Carefully bend the tab to the left ⅟₃₂ inch. Insert a videocassette and push Play. See if the picture has cleared up.

3. If not, then move the tab over another ⅟₃₂ inch and check the picture again.

4. If the picture doesn't clear up after moving the tab over ⅛ inch, then move the tab back into its original position and replace the tension spring.

FIGURE 11.10

Second type of brake shoe band adjustment.

FIGURE 11.11

Using needle nose pliers to adjust the brake shoe band.

Fourth Type

In newer models, at the base of the brake shoe band is a plastic screw head, as shown in Figure 11.12.

1. Simply place a Phillips or flat head screwdriver into the screw head and turn it clockwise ¹⁄₃₂ inch.

2. Insert a videocassette and push Play.

3. See if the picture is clearer. If not, then turn the screw another ¹⁄₃₂ inch and check the picture again.

Remember, after you get a good picture, check the video tape at the take-up tape guide to be sure that the tape isn't being crinkled and eaten. If so, you overtightened this adjustment and you need to loosen it. Refer to Figure 3.1 to orient yourself with the terminology of the videotape path.

Removing a Tension Guide Bracket

First, remove the cassette carriage. Refer to Chapter 8, "Removing and servicing cassette carriages." In older models, refer to Chapter 9,

Fourth type of brake shoe band adjustment.

"Getting into the undercarriage." In newer models that have a mother-board under the transport, you will need to remove the transport. Refer to the section in Chapter 9, "Transports." When removing the tension guide pole bracket, you'll find a plastic arm with a metal or plastic band attached to it. This band wraps around the supply spindle and is called the *brake shoe band* (see Figure 11.8). At the opposite end of the band is another adjustment screw. Do not remove this screw. Read the following sections and follow the instructions that pertain to the model you have.

First Type

In older models:

1. Remove the C-, or E-ring, on the shaft holding the bracket on (Figure 11.13). Refer to the sections in Chapter 7, "Removing C-, or E-rings" under the subhead "Finding the lubricating points."
2. Pull the bracket straight up and off.
3. You'll need to remove the tension guide spring.
4. Holding the bracket in your hand, slip the hook on the end of the spring off the notch on the bracket, as shown in Figure 11.14.

The C-ring holding the back tension guide bracket.

Removing the tension spring.

Second Type

On other models:

1. Remove the tension guide spring first by slipping the spring off the notch.

2. The base of the bracket has a white plastic mount. Pull straight up on the plastic mount for removal. This type of mount snaps on and off.

Third Type

Some models have a latch on the side of the tension bracket, as shown in Figure 11.15.

1. Pull straight back on the latch and, at the same time, pull the bracket straight up and out. The shaft and the tension bracket will come off. The tension spring can be removed before or after removing the bracket.

FIGURE 11.15

A latch holding the back tension guide pole bracket.

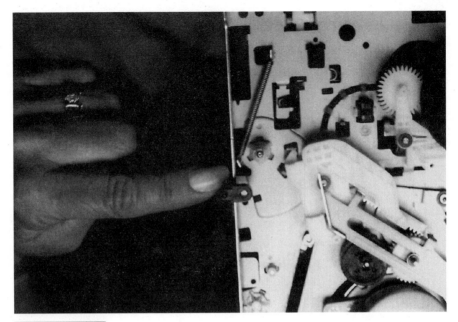

FIGURE 11.16

The tab on the back tension guide pole and the tension spring in the undercarriage.

Fourth Type

In newer models, remove the tension guide spring first. In some units, the spring is connected to the tension bracket by an arm protruding through a hole in the transport to the undercarriage, as shown in Figure 11.16. In most units, the tension spring is mounted on top of the transport. To remove the tension guide bracket, you need to remove the transport. Locate the shaft on the tension guide bracket. The shaft is located right where the bracket pivots. Turn the transport over and look for the shaft. You will see a white or black plastic clip holding the shaft down. There will be one clip on each side of the shaft, as shown in Figure 11.17. Pull both clips back to clear the notch in the shaft and pull the tension bracket straight out.

Disconnecting the Brake Shoe Band

The band has to be disconnected to remove the tension guide pole bracket. At the opposite end of the band, you'll find another adjustment

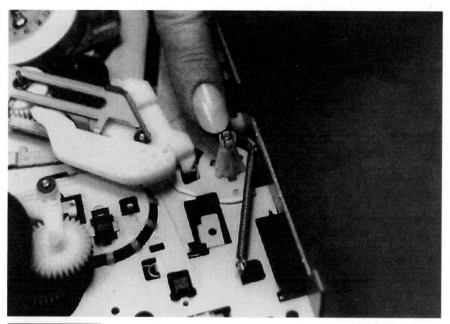

FIGURE 11.17

The clip holding the back tension guide pole bracket in the undercarriage.

screw. Do not remove this screw. This adjustment screw is very crucial. You could mess up the torque adjustment if you turn this screw.

Older models have two ways to remove the band:

1. One type of unit has a small Phillips head screw located on top of the brake shoe arm. By removing this screw, the band will lift right off. Figure 11.18 shows where this screw is located.

2. The second type of unit has a C- or E-ring located on the backside of the tension guide bracket, as shown in Figure 11.19. Just remove the C- or E-ring and the band will come off.

Newer models have two ways to remove the band:

1. In the first type, take hold of the plastic mount on the end of the brake shoe band and with the other hand take hold of the tension guide pole bracket and twist it about 45 degrees and slip them apart. The tabs holding the two pieces together are shown in Figure 11.20. (I removed these pieces for a better view.)

FIGURE 11.18

Where the screw would be located on the brake shoe arm.

FIGURE 11.19

The C-ring located on the backside of the tension guide bracket.

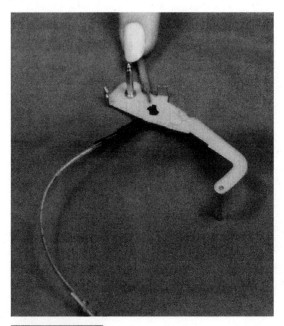

FIGURE 11.20

The tab on the backside of the tension guide pole bracket.

2. In the second type, look on the backside of the tension guide pole bracket for a double clip. Squeeze the clip together and pull the two pieces apart. (I have removed these pieces to give you a better view, as shown in Figure 11.21.)

Repairing the Brake Shoe Band

The brake shoe band is wrapped around the supply spindle with felt inside, (see Figure 11.8). Sometimes the felt will separate from the metal or plastic band and you can see the felt sticking out from behind the band. This will cause less pressure on the supply spindle that causes less drag; less drag causes less torque on the videotape.

To repair these problems, remove the cassette carriage:

1. Remove the band, as in the previous paragraph.

2. Place some super glue on the inside of the plastic or metal band. Be sure to use very little glue because if you use too much the glue will soak into the felt and make the felt hard, causing too little drag.

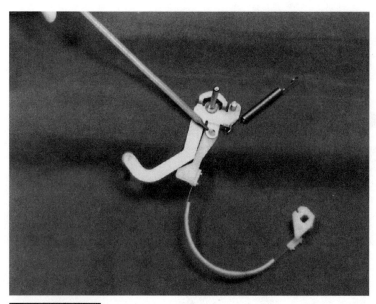

FIGURE 11.21

The clip on the backside of

To straighten out the tension bracket arm:

1. Place the bottom side of the arm onto a hard surface, allowing the shaft on the bracket to hang over the edge of the hard surface.

2. With a hammer, very carefully tap on top of the arm until the bracket lies perfectly flat onto the flat surface.

3. To remount the straightened bracket, simply reverse the procedure. If you are unable to straighten the bracket out, you will need to replace the bracket. To replace the tension guide pole bracket, refer to the section in Chapter 1, "Parts."

3. Carefully place the felt back onto the band. Be sure that the felt is aligned with the sides of the band.

4. Hold the felt in place until the glue has set.

If you are unable to repair the brake shoe band, refer to the section in Chapter 1, "Parts."

Repairing a Bent Tension Guide Pole Bracket

If the arm on the tension guide pole bracket is bent, it'll drag on the top of the transport or on some other bracket near the tip of the bracket where the guide pole is mounted. The place where the bracket usually gets bent or twisted is shown in Figure 11.3. In other cases, the bracket arm could be twisted up and not be dragging on anything, but will cause the guide pole to be tilted. This misalignment causes the wrong amount of torque on the videotape.

Review

1. First determine if the problem is a dirty video head, misaligned roller guides, a misaligned A/C head, or a misaligned tension guide.

2. To determine which problem it is, place your finger on top of the tension guide pole and move it to the left and to the right slowly, periodically stopping to see if the picture clears up.

3. If the picture clears up, the back tension guide is out of adjustment. If you can't remove the stationary lines, the roller guides are out of alignment. If the picture stays completely distorted or snowy, then the video head is dirty. If the picture continues pulsating, then the A/C head is out of alignment.

4. When tightening the spring adjustment, move it notch by notch until the picture clears up.

5. Do not overtighten the spring!

6. Be sure that the back tension guide moves freely before making this adjustment.

7. Only make the brake shoe band adjustment after making the spring adjustment or there is no spring adjustment.

8. Brake shoe band adjustment is very crucial, make very small adjustments until the picture clears up.

Video Tape Path Alignment

Videotape path alignment is the most common repair problem. I highly recommend you read this chapter carefully before performing this adjustment.

Detecting Misaligned Roller Guides

Connect the TV monitor to the VCR and turn it on. Insert a commercial videocassette. Push Play. You'll notice a horizontal belt of lines across the screen. This belt of lines is in either the upper or lower portion of the TV screen, as shown in Figure 12.1. You might have two or more belts of lines across the TV screen, as shown in Figure 12.2. You might find that ⅔ of the picture is washed out and distorted in either the upper or lower portion. Locate the tracking adjustment on the face panel of the VCR or on your remote control. The tracking adjustment is used to remove the lines from the picture. When you try to remove the lines, you'll notice that they either move up or down the screen, but still remain.

FIGURE 12.1

A belt of lines across the bottom portion of the screen.

FIGURE 12.2

Two belts of lines across the bottom portion of the screen.

Another problem occurs when someone lends you a videocassette they've recorded and lines appear upon playback in your unit. To determine if your machine or theirs is out of alignment, place the videocassette in a third unit. If it has lines in playback, then the bad machine is the other person's. If the third machine doesn't have lines, then the bad machine is yours.

Setting the Tracking Adjustment

To properly perform the necessary adjustment, the tracking control must be centered.

In other models, the tracking adjustment is a knob with a click position. Rotate the knob until you hear the click and leave it in the center position.

In other models, the adjustment is either a dial or a slide control. As you move the control, you'll feel an indentation or niche that is the center position. Set the control to the center position.

In newer models, buttons on the front of the unit or on the remote control are the adjustment. To center the tracking, just insert the videocassette and push Play. The VCR will automatically center itself.

In other models with auto tracking, after inserting the videocassette and it goes into the Play mode, your TV screen will display "auto tracking." Don't start the alignment until the display goes off. Sometimes the auto tracking display will come on while you are aligning the videotape. Stop until the display goes off, and then continue with the alignment. After completing the alignment, push Stop and then Play so that the auto tracking will come back on. This will reset the tracking and if there is any distortion or lines at the top or bottom of the screen go back and touch up the roller guides.

The Terms Load and Unload

A tape is loaded if you have pushed Play and the unit has pulled the videotape out of the videocassette and onto the videotape path. Videotape is unloaded if it has been inserted into the unit and the videotape is still in the videocassette. Keep these terms in mind; you'll be reading about them frequently.

Identifying Roller Guides and V-mounts

The supply roller guide is on the left side of the video drum as you face the front of the unit. The other roller guide is the take-up roller guide, which is on the right side of the video drum, as shown in Figure 12.3.

Each roller guide is adjustable by turning the post. The post is the part of the guide with the white glass roller on it. The glass roller is the part that the videotape comes into contact with. All roller guideposts have a fitting on the top, which a roller guide alignment tool or Allen wrench fits into. (Look at the top of the post in Figure 12.6. You can see slots on each side of the shaft.) To make these adjustments, you'll need a 1.5-mm, a 1.27-mm, or a 0.89-mm Allen wrench. Most units need a tape guidepost tool.

You have two options: take a thin-blade flathead screwdriver and file a notch in the center of the blade or purchase a tape guidepost tool. Refer to the sections in Chapter 1, "Making a slotted screwdriver" or "Parts."

FIGURE 12.3

Supply roller guide on the left and take-up roller guide on the right.

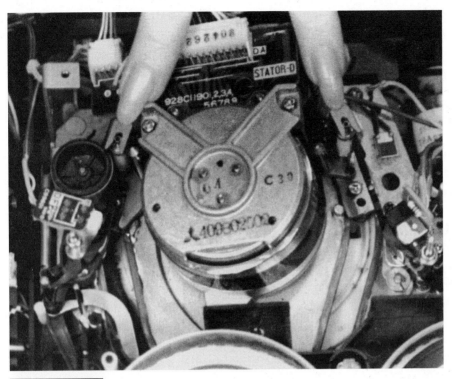

FIGURE 12.4

The V-mounts.

The first purpose of the roller guides is to load the videotape onto the videotape path. The second purpose of the roller guides is to keep the videotape aligned to the video drum so that the video heads pick up the correct image.

Load the unit and push Play so that you can see the roller guides in their locked positions. Notice the roller guides are locked into position by V-mounts, one on each side of the video drum, as shown in Figure 12.4. To make these adjustments, the roller guides must be in the locked V-mount position.

Loosening the Base Lock Nuts

All roller guides are held by a lock nut that has to be loosened to adjust the roller guides. You need to use a 1.5-mm, 1.27-mm, or 0.89-mm Allen

If you insert a videocassette and the unit goes into Fast Forward or Rewind instead of Play, it's caused by too much room light. You'll need to prepare the unit before you can make these adjustments:

1. Locate the cassette holder in the carriage.

2. On each side of the cassette holder is a small round hole, approximately an ⅛ to a ¼ of an inch in diameter (refer to Figure 21.4). When a videocassette is inserted, these holes line up with the holes on the tape sensors. Take a piece of black electrical tape and place a small piece over each hole.

3. Another option is to shade the light so that the lighting does not shine directly on the VCR.

wrench or a small-tipped Phillips screwdriver to loosen these nuts. They must be loosened before the roller guides are loaded on their V-mounts. Turn the nut counterclockwise ½ turn to loosen the roller guides. Read the following locations of lock nuts and choose the one that fits your unit.

First Location

In most models, the lock nut is located on the back side on the base of the roller guide post or on the brass base itself, as shown in Figure 12.5. If you have a unit with a brass base roller guide, hold each roller guide and pull up. If the roller guide lifts up, then the post is broken. Proceed to the section in this chapter, "Broken roller guide post."

Second Location

In some models, the lock nut is located on the side of the base of the roller guidepost, as shown in Figure 12.6.

Third Location

In other models, the V-mount is connected to the base of the roller guidepost and the lock nut is located under the V-mount, as shown in Figure 12.7.

Fourth Location

In a few models, the lock nut is located on top of the roller guide base, directly in front of the roller guidepost, as shown in Figure 12.8. A Phillips head screw is directly in front of the lock nut. Do not touch it. Insert a videocassette and push Play. This type of model must be in the loaded position in order to turn the lock nuts.

FIGURE 12.5

First location for a lock-nut.

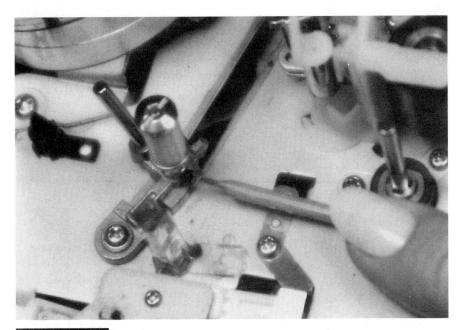

FIGURE 12.6

Second location for a lock-nut.

FIGURE 12.7

Third location for a lock-nut.

FIGURE 12.8

Fourth location for a lock-nut.

Broken Roller Guide Post

Roller guideposts only break in models with a brass base roller guidepost (refer to Figure 12.5). The brass base is compressed into the roller guide mount, the bonding breaks, and the post slips down.

Adjusting the Base Lock Nut

Regardless of the type of lock-down system in your unit, follow these next steps:

1. Place either an Allen wrench or filed screwdriver on top of the roller guidepost. You should be able to turn each roller guide back and forth smoothly. If it feels a little sticky, go back and loosen the base lock nut ¼ turn. Recheck the smoothness of movement of each roller guidepost. It's important that the roller guides turn smoothly.

2. After checking the roller guideposts for smoothness of movement, insert a videotape and push Play.

3. Look carefully at the top of each roller guidepost. The torque on the video-tape might cause the guideposts to spin or turn. If either one does, tighten the base lock nut slightly. This adjustment is extremely temperamental. If the adjustment is too loose, the roller guide will go out of alignment after you remove the tool.

QUICK REPAIR

1. **Pull up on the post. Look for a fine line caused by tarnish to find the post's original position. You also could look at the other roller guide and position it to the same height. The top of the roller guides should be the same height. Eyeballing this adjustment is fine; it doesn't have to be exact.**

2. **Be sure that the lock nut is pointing straight back, approximately 1 mm above the roller guide mount. Once it is set, don't touch it.**

3. **Run a bead of super glue around the base of the roller guidepost. Avoid placing any glue on the lock nut. Allow the glue to dry until the piece is set.**

QUICK»TIP

There's one more crucial point (refer to Figure 12.8). Units with a top base lock nut also have an angle nut adjustment behind the V-mount, as shown in Figure 12.9. Do not confuse this nut with the lock nut. The angle nut is factory set. Under no circumstances should it be turned.

The angle nut adjustment.

Roller Guide Functions

The roller guides pull the videotape out of the videocassette onto the videotape path. When the unit is in the Play mode, the supply roller guide aligns the videotape to the left side of the video drum. The take-up roller guide aligns the videotape to the right side of the video drum. Both roller guides align the videotape to the video heads.

Adjusting the Roller Guides

On the lower video drum is a tape edge guideline, as shown in Figure 12.10. Notice that the slightly protruding edge runs on a downward angle from left to right.

1. Adjust the roller guides so that the bottom edge of the videotape sits on the tape edge guideline.

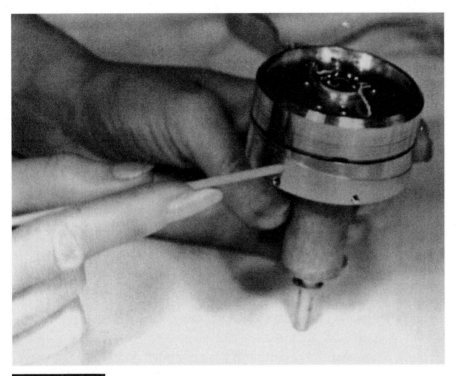

FIGURE 12.10

Tape edge guideline.

2. Push Play and load the videotape. Look down at the lower video drum. Look for a slightly protruding shiny edge at the bottom of the videotape, as shown in Figure 12.11. The shiny protruding edge should be visible all the way between the two roller guides.

3. The bottom of the videotape should just touch the top of the tape edge guideline. If you don't see this edge all the way around, locate the roller guidepost that is closest to where the video tape covers the edge and slowly turn it counterclockwise.

4. This procedure will move the videotape further up the video drum. When the shiny edge is exposed, stop turning the adjustment.

FIGURE 12.11

The protruding edge at the bottom of the video tape.

Now, align each roller guide to the tape edge guideline. Start with the take-up roller guidepost, the one on the right side of the video drum:

1. Insert the proper tool and turn it clockwise. This adjustment moves the videotape down.

2. As the tape moves down, keep an eye on the shiny protruding edge and on the bottom edge of the videotape. As you slowly turn the roller guide clockwise, at some point, you'll see the video tape wrinkle. These wrinkles flow with the videotape around the video drum and toward the take-up roller guide, starting from a point 1 to 1½ inches away, as shown in Figure 12.12. If you turn the adjustment too fast, the videotape will pop over the shiny protruding edge instead of wrinkling.

3. When you start to see the wrinkle, stop turning the adjustment. Now, turn the adjustment counterclockwise until the wrinkle disappears. Turn it in the same direction for another ¼ turn and stop.

The guide now is properly set. Follow the same procedure with the supply roller guide, the one on the left side of the video drum. The

FIGURE 12.12

Wrinkling of the video tape at the protruding edge.

only difference you'll find is that the wrinkles at the bottom of the videotape move away from the roller guide, rather than toward it.

Checking the Adjustment

Connect the VCR to the TV monitor and check the adjustment of the roller guides. Insert a high-quality rented or purchased videotape, which has been produced by a well-known production company. Now, push Play. If the roller guides are prop-erly adjusted to the tape edge guideline, the picture will be perfect. If not, you might experience one of two things. A thin belt of lines might run horizontally across the top or bottom of the picture. On the other hand, the picture might jump up and down or jitter. If either of these problems occur, fine-tune the roller guides. If you have an auto track-ing VCR, refer to the section in this chapter, "Setting the tracking ad-justment."

QUICK»»TIP

You can save some time when adjusting the roller guides by determining which guide needs to be realigned.

1. Push Play to load the videotape.

2. While the unit is running, try to turn each roller guidepost.

3. If one post can be turned easily while the other can't, you should only need to realign the one that you can turn.

Fine-tuning the Roller Guides

When fine-tuning the roller guides, use only a high-quality videotape. If you have any lines across the bottom of the picture, as shown in Figure 12.13, insert the proper tool into the take-up roller guide and turn it clockwise until all the lines disappear from the bottom of the screen. When you reach this point, stop and make a mental note of this position. Then, continue in the same direction for another ¼ turn and stop. Remove the tool. The take-up roller guide is now fine-tuned.

If any lines run across the top of the picture, as shown in Figure 12.14, insert the proper tool into the supply roller guide and turn it clockwise until all lines disappear off the top of the screen. When you reach this point, stop and make a mental note of this position. Then, continue in the same direction for another ¼ turn and stop. Remove the tool. The supply roller guide is now fine-tuned.

When fine-tuning the tape-path alignment and more lines appear when turning the roller guide clockwise, the positioning of the video-

FIGURE 12.13

A line across the bottom of the picture.

FIGURE 12.14

A line across the top of the picture.

tape edge guideline is incorrect. Go back to the section in this chapter on "Adjusting the roller guides."

If, after completing the alignment, you find a confusing number of lines, then both roller guides are out of alignment. Return to the tape edge guideline adjustment and readjust from that point.

These adjustment procedures are correct for all two-hour videocassettes if you are using a professionally produced videotape.

Fine-tuning the Roller Guides for Six-hour Tapes

When inserting a six-hour videocassette after you've fine-tuned for a two-hour cassette, you might find a thin line or

QUICK REPAIR

You might find jittering in the picture after you've adjusted the videotape to the tape-edge guideline or after you've fine-tuned it. To correct this problem:

1. Insert the proper tool into the supply roller guide and turn it clockwise approximately ¼ turn.

2. Fine-tune the adjustment back and forth, not going over ⅛ turn, until you find where the jittering stops.

3. If this adjustment doesn't completely eliminate the jittering, you can fine-tune the take-up guide. You shouldn't have to turn the take-up more than ⅛ turn. This final adjustment will correct the problem.

FIGURE 12.15

Fine lines in the picture.

lines in the upper or lower portion of the picture, as shown in Figure 12.15. In this case, you need to fine-tune the six-hour tape. Insert a six-hour recorded videotape. Try to find a tape that was recorded by your machine when you first purchased it. The unit was properly aligned at that time; consequently, the videotapes also will be aligned. Simply fine-tune the roller guides, as previously explained.

After aligning the roller guides and inserting a video cassette recorded just prior to the alignment procedure, you'll probably find that the tape now has lines in the picture, but all older video cassettes function properly. Any previously recorded tapes (with lines in the picture) were recorded when the unit was out of alignment.

Tightening the Base Lock Nuts

The last step is to tighten the lock nuts on the roller guides. Locate the roller guide lock nuts, insert the proper tool, and turn it clockwise until the nut is snug. Do not overtighten the nut. Overtightening can cause the roller guide to turn and you'll lose the alignment.

Review

1. Center the tracking control.

2. Loosen the base lock nut before turning the roller guides.

3. Be sure that the roller guides turn smoothly.

4. Be sure that you don't turn the base lock nut too far clockwise or the roller guides will turn by themselves as the videotape passes around them.

5. Don't touch the angle adjustment behind the V-mount.

6. Adjust the bottom edge of the videotape to sit on the tape edge guideline.

7. Fine-tune the roller guides (the take-up roller guide affects the bottom of the picture and the supply roller guide affects the top of the picture).

8. Tighten up the base lock nuts.

Audio Head Alignment

The audio head is incorporated into the A/C head. Usually a dirty or improperly aligned A/C head causes a weak or distorted audio signal. A buildup of residue prevents the videotape from making flush contact with the A/C head, producing a weak or distorted audio signal. Refer to the sections in Chapter 4, "A/C head and erase head" under the subhead "Cleaning the various components." If you have a stereo hi-fi model, you'll have to put the stereo hi-fi switch in the Normal position before proceeding with this alignment. If you have a stereo switch, place it into the Stereo position.

Ideal Setting for an A/C Head

Figure 13.1 is a diagram of the front of an A/C head. It has three sets of fine parallel lines or small metallic rectangular squares. Beginning with the top left side, the first metallic square is the audio erase head. Straight across, on the upper right side, is the audio playback and record head. The third metallic square, the FM head, is at the bottom right side. The distance between the center of the audio head and the

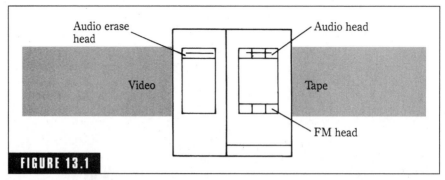

FIGURE 13.1

Diagram of an A/C head.

center of the FM head is the exact width of videotape. The ideal setting for an A/C head is straight up and down, not tilted. The top edge of the videotape should barely touch the centerline of the audio and erase head, as shown in Figure 13.2. In some models, you'll notice a little screw and shield plate mounted on top of the A/C head, as shown in Figure 13.3. Remove the shield plate so that you have a clear view of the front of the A/C head.

FIGURE 13.2

The ideal setting for an A/C head.

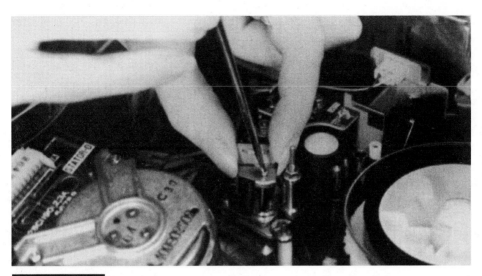

FIGURE 13.3
The shield plate and its screw.

Location of Adjustment Screws

All A/C heads are mounted on a set of brackets, as shown in Figure 13.4. The upper bracket has three adjustment screws and fastens directly to the A/C head. The three adjustment screws pass through the upper bracket to the lower bracket. The upper bracket is mounted to the lower bracket by these three screws making the upper bracket ad-

FIGURE 13.4
The A/C head mounting brackets.

justable. Read each of the following sections and find the bracket that resembles the A/C head bracket in your VCR.

First Location

In older models, the A/C head adjustment screws are located on each side and behind the A/C head. Figure 13.5 shows the upper bracket and provides a better view of the three adjustment screw holes. When this bracket is mounted in its proper position, you'll find an adjustable setscrew going through each screw hole. The spring-loaded screw is located on the left side of the A/C head. The tilt adjustment is directly behind the A/C head. The azimuth adjustment is on the right side of the A/C head.

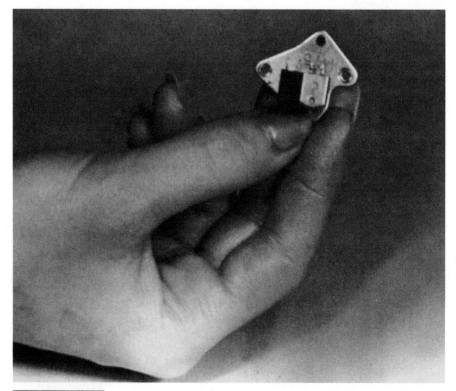

FIGURE 13.5

The three adjustment screw holes.

FIGURE 13.6

The azimuth adjustment.

Second Location

In other models, the A/C head adjustment screws are located behind the A/C head. The spring-loaded screw is on the right side and moved further behind the A/C head. The tilt adjustment is still directly behind the A/C head. The azimuth adjustment is on the right and behind the A/C head, as shown in Figure 13.6.

Third Location

In newer models, the A/C head-adjustment screws are located in front and behind the A/C head, as shown in Figure 13.7. This unit has two springs. One spring is located between the brackets under the azimuth adjustment screw on the left side of the A/C head. The other spring is located at the rear on a sliding post between the brackets. The tilt adjustment is in front of the A/C head. The adjustment screw to the right and behind the A/C head is part of the height adjustment. If you have this type of VCR, read the section in this chapter entitled "Spring-loaded A/C head."

Fourth Location

In other new models, all three adjustment screws are spring loaded. The springs are located between the upper and lower brackets, as

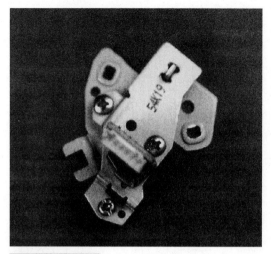

FIGURE 13.7

Different locations of the adjustment screws.

shown in Figure 13.8. (I've picked the A/C head and brackets up to give you a better view.) The tilt adjustment is directly behind the A/C head (Figure 13.8) or directly in front of the A/C head, as shown in Figure 13.9. The azimuth adjustment is on the left side of the A/C head. The same adjustment screws are used for another purpose. If you have this type of VCR, read the section in this chapter entitled "Spring-loaded A/C head."

Function of Adjustment Screws

Locate the setscrew on the right side of the A/C head with a spring wrapped around the screw above the bracket, as shown in Figure 13.10. This spring-loaded Phillips screw keeps tension on the A/C head. You shouldn't have to adjust this screw unless you're replacing the A/C head. If you replace the head, remount the screw with the spring on it. Adjust the screw down until there is tension on the A/C head, but the spring will not be all the way collapsed (see Figure 13.10). When adjusted properly, the A/C head is stiff, but remains flexible.

The second setscrew is located directly behind or directly in front of the A/C head and usually will have a 1.5-mm Allen nut or a Phillips head screw. This screw is the tilt adjustment, which allows you to change the angle of the A/C head forward or backward.

An A/C head mounted on spring loaded screws.

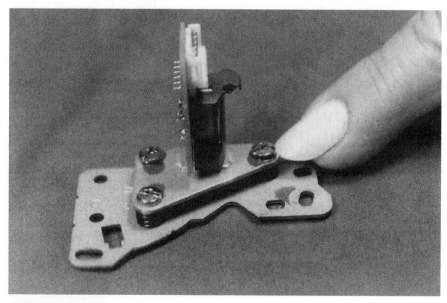

The tilt adjustment.

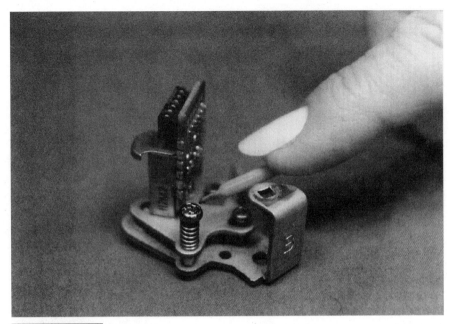

The tension screw adjustment.

The third setscrew is on the left side of the A/C head and is a Phillips head screw. This screw is for making the azimuth adjustment, which allows you to change the angle of the A/C head from side to side.

The last adjustment is directly behind the A/C head. This nut adjustment controls the height of the A/C head. You'll find a metal bracket or shaft with a nut on top of it directly connected to the bottom bracket, as shown in Figure 13.11. This adjustment nut moves the A/C head and both brackets up or down.

To adjust this nut, you'll need one of these tools:

■ $\frac{5}{16}$-inch socket (or nut driver)

■ $\frac{7}{32}$-inch socket (or nut driver)

■ ¼-inch socket (or nut driver)

For a few models, use a filed flathead screwdriver (refer to Figure 1.6)

The height adjustment.

Tilt Adjustment

Refer to Figure 3.1 to orientate yourself with the terminology of the videotape path. Insert a videocassette and push Play. Look straight down on the videotape to see if it looks like a perfectly straight hairline between the take-up roller guide and the A/C head. Figure 13.12 shows you where to look. Also check to see if the videotape is being wrinkled at the bottom protruding edge of the take-up tape guide, as shown in Figure 13.13. When the A/C head is adjusted properly, the videotape on the take-up tape guide will be running smooth.

If the video tape has a twist coming from the roller guide or is being wrinkled at the take-up tape guide, place a 1.5-mm Allen wrench or a Phillips head screwdriver into the adjustment screw and slowly turn it counterclockwise until the twist or wrinkle disappears. Then, turn the adjustment clockwise to move the top of the A/C head forward toward the videotape. At the same time, look straight down on the videotape where it comes off the take-up roller guide and makes contact with the A/C head. Be sure that the videotape doesn't get twisted. If you go too far with this adjustment, it causes the videotape to wrinkle on the bottom protruding edge of the tape guide. If this happens,

FIGURE 13.12

The point to look straight down at the video tape.

FIGURE 13.13

The protruding edge on the take-up tape guide.

slowly turn the adjustment counterclockwise until the wrinkle disappears. You want to move the A/C head as far forward as possible without leaving a twist in the video tape. If the twist still remains, tighten the torque adjustment. Refer to the sections in Chapter 11 entitled "Back tension guide adjustment" and "Brake shoe band adjustment."

Height Adjustment

Locate the height adjustment nut on top of the metal bracket or shaft directly behind the A/C head. Find the proper size of socket or nut driver and place it on the nut. Turn the nut counterclockwise while keeping an eye on the front of the A/C head. Watch for the metallic square on the face of the A/C head to come up above the videotape, as shown in Figure 13.14. If you have the TV monitor on at this time, the picture will have a belt of horizontal lines running up through it. Do not worry, the lines will disappear after finishing the audio alignment. Proceed to the azimuth adjustment.

FIGURE 13.14

Adjusting the metallic squares above the video tape.

Azimuth Adjustment

Insert a Phillips head screwdriver into the azimuth adjustment. This adjustment tilts the A/C head from side to side. Adjust the A/C head so that the top two metallic squares are horizontally parallel to the top edge of the videotape. These two metallic squares are the audio eraser head and the audio playback head.

Now, go back to the height-adjustment nut. Turn the nut clockwise. Move the A/C head down until the top edge of the videotape just covers the centerline in the middle of the two metallic squares. Figure 13-2 shows how the videotape should look up against the A/C head.

Fine-tuning the A/C Head

You'll need to fine-tune the audio head once the other adjustments are completed. Insert a commercial videotape. Approximately thirty seconds to five minutes after the movie has ended, the monitor will have a white raster or a white blank screen. During this time, you'll hear a 1000-note cycle audio tone that lasts up to five minutes. This audio tone is a wake-up call for anyone who fell asleep during the movie. You're going to fine-tune the A/C head using the audio tone.

1. Locate the audio signal on the tape and turn up the audio on the TV monitor.

2. Fine-tune the height adjustment by turning the nut back and forth slightly while listening to the tone.

3. Determine the point at which the volume peaks. Set the adjustment here. You shouldn't have to turn this adjustment more than ¼ turn in either direction.

4. Locate the azimuth adjustment. Use a Phillips head screwdriver to turn the adjustment screw in either direction and listen to the tone. You should hear the audio tone increase or decrease in volume. Be sure to turn the adjustment at a moderate speed. You want to turn the adjustment in the direction that causes the volume to increase.

5. Keep going in this direction until the volume peaks and then the audio signal starts lowering in volume.

6. Keep turning the screw back and forth until you find the loudest point and leave it set there.

If you have a stereo VCR (a stereo unit without stereo hi-fi), place the stereo switch on the face panel into the Stereo position and use the same procedure for fine-tuning the A/C head, except in this case, listen to both speakers on your TV stereo monitor. When peaking the height and azimuth adjustments, be sure that both speakers are equal in volume.

> **QUICK»TIP**
>
> You'll usually find red or green paint on the side of the adjustment screws. The paint locks the screws down, keeping the screws from turning on their own. After aligning the A/C head, place red nail polish on the side of these screws to lock them down.

Spring-loaded A/C Head

These models have three setscrews, but no height-adjustment nut. All setscrews are located in the same places; the difference is that they're all spring-loaded, (see Figure 13.8 and 13.9). Adjust the height of the A/C head by turning all three screws, one after the other. Turn them counterclockwise to move the head up or clockwise to move the head down. Start with any screw and count each turn you make. Turn the other two screws the same number of turns as the first to keep the audio head straight up and down.

Turning the screws counterclockwise causes the metallic squares to rise on the face of the A/C head above the top edge of the videotape. As you bring the head back down, you'll need to adjust each screw accordingly. Keep the top edge of both metallic squares parallel to the top edge of the videotape. Adjust the metallic squares down to the centerlines, as previously explained.

After you set the height, read the sections in this chapter "Third and fourth location" under the subhead "Locating the adjustment screws." Then, read "Tilt adjustment" and "Fine-tuning the A/C head." Although this method is harder to align, it maintains the alignment longer.

A Tip on Recording Bad Audio

Upon playing back a personal recorded videocassette, the picture is okay, but the audio portion is distorted or weak. A dirty A/C head could cause this condition. The head was just dirty enough to keep the audio signal from imprinting properly on the videotape. On other prerecorded tapes, the audio signal is okay. Clean the A/C head and record the tape again.

Adjusting the FM Portion

If you have lines floating from the bottom to the top of the picture, the quickest way to correct this problem is to adjust the height adjustment. Adjust the height adjustment down below the center line, then move it up until the top edge of the video tape just covers the center line in the top metallic boxes on the front of the A/C head (see Figure 13.1). The problem is that the videotape is either too high or too low on the A/C head, so the FM portion of the A/C head is unable to read the proper signal off the videotape. (Refer to Chapter 14, "FM alignment.")

Review

1. Clean the A/C head.

2. There are four audio head adjustments.
   A. The setscrew with a spring wrapped around the screw puts tension on the A/C head.
   B. The screw directly behind or in front of the A/C head, called the *tilt adjustment*, is for taking wrinkles and twists out of the videotape.
   C. The screw on the opposite side of the spring-loaded screw is the azimuth adjustment, for peaking out the audio signal.
   D. The nut on top of the bracket or shaft behind the A/C head is the height adjustment and is used to line the top edge of the video tape to the center line in each metallic square located near the top of the A/C head.

3. Adjust the tilt adjustment to allow the most pressure against the videotape without twisting or wrinkling the tape.

4. Adjust the height adjustment so that you can see the metallic square on the face of the A/C head above the videotape.

5. Adjust the azimuth adjustment to make the metallic squares on the A/C head parallel to the top of the videotape.

6. Adjust the height adjustment down to where the top edge of the videotape is just covering the center of both metallic squares and using an audio tone for peaking out the audio signal.

7. Adjust the azimuth adjustment using an audio tone for peaking out the audio signal.

8. In spring-loaded units, use all three adjustment screws for adjusting the height by counting each turn of each screw.

FM Alignment

Videotape carries three separate signals. Figure 14.1 shows where the signals are put on videotape. The top portion of the videotape carries the audio signal, which is adjusted by doing an audio head alignment. The middle portion carries the video signal, which is adjusted by doing a video tape path alignment. The video signal is produced on tracks. The bottom portion of the videotape carries the FM signal.

Checking the Tracking Control

Older models have the tracking-control knobs on the front of the unit. Insert videotape and push Play. Place the tracking control in its center position. It should have an indentation or a niche. Rotate or slide the tracking-control knob to the right until lines appear either from

> **QUICK>>TIP**
>
> The FM alignment centers the tracking control. If the audio and video portions are functioning properly (no weak audio or no stationary lines in the picture), you can proceed with this adjustment. The FM alignment should be done after an audio head alignment or a videotape path alignment has been completed.

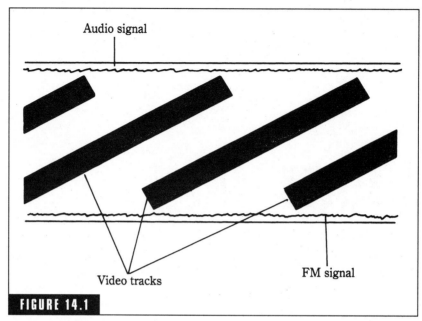

FIGURE 14.1

The three signals on a video tape.

the top or the bottom of the picture or until the picture becomes completely distorted with lines. Return the tracking control to the center position and note the distance you traveled to the right before the lines appeared. Now turn or slide the tracking-control knob to the left until the lines or distortion appear in the picture, again noting the distance back to the center position. You need to see if all the lines start at the same point in either direction from the center.

If the adjustment isn't centered properly, you'll be able to adjust the tracking control all the way in one direction without showing any lines from the center point, but in the opposite direction, lines appear right away. On the other hand, you might find that the tracking control is set at one end or the other in order to get a clear picture. To correct these problems, perform an FM alignment.

Newer models have adjustment buttons on the front of the unit or on the remote control. When you insert the videocassette and push Play, the VCR will automatically center itself. Push on either button until lines appear in the picture or until the picture becomes completely distorted with lines. Note the time it takes. Next, push Eject and re-insert the videotape. This will automatically re-center the tracking. Now, push on the other button until lines appear or until the picture becomes completely distorted. Note the time again. You need to see if it takes the same amount of time to reach the lines or the picture has the same amount of lines or distortion when you push on either button.

If the buttons aren't centered properly, you'll be able to push one button without any change, but when you push the other button, the picture will become completely distorted. On the other hand, you might find that you have to push one of the two buttons to clear up the picture when you first insert videotape. If that happens, you need an FM alignment.

In newer models with auto tracking, after inserting the videocassette and it goes into Play, your TV screen will have a display saying "auto tracking." Wait until the display goes off. If any distortion or lines are at the top or bottom of the screen, do an FM alignment.

Locating the FM Adjustment

In some models, the FM-alignment adjustment is a nut and can be located at the front right corner of the A/C head mounting bracket, as shown in Figure 14.2. In a few models, this adjustment is located to the left side of the A/C head. In other models, the adjustment is located directly behind or behind and to the left of the A/C head, as shown in Figure 14.3. The FM alignment adjustment nut will be either a large brass or silver adjustment nut with slots on top and on each side of the threaded shaft where it's mounted. To adjust this nut, place a medium flathead screwdriver in either slot, as shown in Figure 14.4, or you could use a slotted screwdriver. On the other hand, you could purchase an audio and control head tool (refer to the sections in Chapter 1, "Making a slotted screwdriver" or "Parts").

Other models have no adjustment nut. You will find an adjustable lever with a slot in it. The V-mount bracket has another slot, as shown in Figure 14.5. This adjustable lever might be located at the rear of the A/C head or to the right side of the A/C head. This lever is attached to the lower A/C head bracket. On top of the adjustable lever is a lock screw. To adjust this lever, loosen the locking screw and place a medium flathead screwdriver in both slots and twist the screwdriver to adjust the A/C head.

FIGURE 14.2

The FM adjustment nut in front of the A/C head.

FIGURE 14.3

The FM adjustment nut behind the A/C head.

FIGURE 14.4

Adjusting the FM alignment nut.

FIGURE 14.5

The FM adjustment slot and locking screw.

The FM adjustment tab on the top of the housing.

A few models had an adjustable tab on top of the loading motor housing, as shown in Figure 14.6. This tab is connected to the lower A/C head bracket. To adjust this tab, place a screwdriver alongside the tab and twist it to bend the tab forward or backward.

In newer models, the A/C head is mounted with spring-loaded screws between the upper and lower A/C head brackets (see Figure 13.9). The lower bracket is mounted directly to the transport by two locking screws. These screws can be located behind the A/C head or one in front and the other behind the A/C head, as shown in Figure 14.7 (I have drawn arrows to point out the screws). To make the adjustment, loosen the mounting screws. Hold the sides of the A/C head and slide the entire assembly from side to side for adjustment.

FIGURE 14.7

Mounting screws to the lower A/C head bracket.

Centering the Tracking Control

In models that have tracking control knobs:

1. Connect the TV monitor and turn it on.

2. Insert a rented videotape and push Play. The tracking control is set at the center position or niche, and the picture is okay.

3. Adjust the tracking control in one direction. You should find no noticeable change or very little change in the picture.

4. Return the tracking control to the center position. Adjust the tracking control in the opposite direction. Lines should appear soon after you start to turn the tracking control, indicating that the tracking control isn't centered.

5. Return the tracking control to its center position.

6. Slowly adjust the tracking-control knob back in the direction showing the lines.

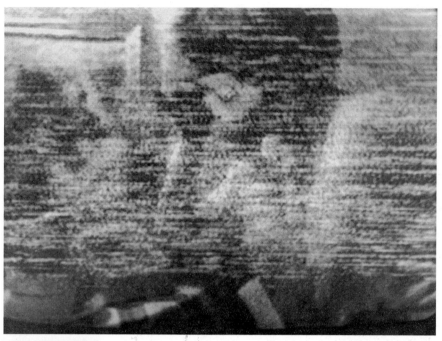

FIGURE 14.8

Lines in the picture.

7. Continue to where three quarters of the picture is covered with lines, as shown in Figure 14.8. Leave the control set at that point.

8. Place an FM-alignment tool or a flathead screwdriver into the FM-alignment adjustment nut and turn it in either direction.

9. As you turn it in one direction, you should notice that the picture is getting worse. When you turn it in the opposite direction, the picture will start to clear up. Continue in this direction until the entire picture is clear and then stop. In most models, it'll take about a three-quarter turn, but in some models, it can take up to two complete turns to remove the lines.

10. Place the tracking control back to its center position.

11. Turn or slide the tracking-control knob in both directions to see if the lines start to appear at the exact same point from the center.

12. If it isn't centered, adjust the tracking control in the direction showing the lines when the control is the closest to the center position or niche.

In models that use push buttons for the tracking adjustment:

1. Insert a rented videocassette and push Play. The VCR will automatically center itself.

2. Push on one button. You should find no noticeable change or very little change in the picture.

3. Push the other button. The picture should become completely distorted, indicating that the tracking control isn't centered. Push the proper button so that three-quarters of the picture is covered with lines (Figure 14.8). Leave the buttons set at that point.

If you have the nut adjustment:

1. Place an alignment tool or screwdriver in the nut (see Figure 14.4). If it has an adjustable bracket, loosen the locking screw and place an alignment tool or a medium-sized flathead screwdriver in the slot (see Figure 14.5). If it has a tab adjustment, place a screwdriver along side of the tab (see Figure 14.6).

2. Turn or twist the screwdriver in one direction. You should notice the picture getting worse. When you turn or twist it in the opposite direction, the picture will become clearer. Continue in this direction until the entire picture is clear and then stop.

3. Tighten any locking screws.

4. Push Eject and re-insert the videotape.

5. Recheck both buttons to see if they have the same amount of lines appearing in the picture from the center point. If not, repeat the same procedure.

If spring-loaded screws are between the upper and lower A/C head brackets (see Figure 13.9):

1. Set the tracking buttons so that three-quarters of the picture is covered with lines.

2. Leave the buttons set at that point.

3. Loosen the mounting screws (see Figure 14.7).

4. Hold the sides of the A/C head and slide the entire assembly in one direction or the other. Some units have gauges so that you can see how far you have moved the A/C head, as shown in Figure 14.9, or the wide notch in front and to the left of the A/C head is used for a gauge (refer to Figure 13.7). A few units have no gauge. The assembly should not move over ½ inch.

5. Slide the assembly slowly in the direction so that the picture starts to become clearer. Continue in this direction until the entire picture is clear and then stop.

6. Tighten the locking screws.

7. Push Eject and re-insert the videotape.

8. Recheck both buttons to see if they have the same amount of lines appearing in the picture from the center point. If not, repeat the same procedure.

FIGURE 14.9

The FM gauge on the lower A/C head bracket.

Tracking Adjustment is Completely at One End

In models that have tracking-control knobs, if the tracking control must be adjusted all the way to one end or the other to receive a clear picture, you need to get the tracking control back to its indentation or niche. Set the control for a good picture; adjust the control back until you receive lines covering three-quarters of the picture. Now, turn the FM adjustment nut to clear the lines, as previously explained.

Repeat this procedure until the picture is clear when the tracking control is in its centered position or niche. You shouldn't need to turn the adjustment more than two or three complete turns. Now, follow the instructions in the previous section to finish centering the tracking control.

In models that use push buttons for the tracking adjustment, you need to push one of the two buttons to clear up the picture when you first insert the videotape. You need to get the tracking control back to its center. Push the proper button until you receive a good picture, then push the other button until you receive lines covering three-quarters of the picture. Move the FM adjustment or slide the complete assembly to clear the lines, as previously explained.

Review

1. Only use a high-quality rented or purchased videotape to make tracking adjustments.

2. Turn the tracking-control knob or push the tracking-control buttons to see if you have lines appearing in the picture at equal distances from the center position.

3. If lines appear in only one direction, you need to adjust the FM adjustment.

4. Turn the tracking-control knob or push the tracking-control buttons until the lines appear in the picture and adjust the FM adjustment until they disappear.

5. In spring-loaded types, loosen the mounting screws and slide the entire assembly until the lines disappear.

6. Recheck the tracking control in each direction for equal distance from the center position when the lines appear in the picture.

Gear Alignment

This chapter deals with gear alignment. All gears, sliding gears or arms and cam gears have an alignment mark. The marking for gears is just behind the teeth on the gear. Look for a hole, arrow, dot, post, or notch. The marking also will be offset from all other marks, being that the gear has any other marks. Each alignment mark will be the only mark of that type on that gear. Do not plug the unit in while doing a gear alignment unless you are told to do so. To do all gear alignments, other than the pinch roller, cassette carriage and take-up gear alignment, you need to remove the transport or get into the undercarriage. Refer to Chapter 9, "Getting into the undercarriage" and the section, "Transports."

When to do a Gear Alignment

1. Remove the top cover, insert the videotape, and push Play.

2. If the roller guides go into their V-mounts (refer to Figure 12.4), but the unit jams up before the pinch roller can clamp up against the capstan shaft, check the pinch-roller alignment.

3. If the roller guides only go half way to the V-mounts before the unit jams up, check the main cam gear for alignment.

4. If only one roller guide goes into its V-mount, check the loading gear alignment.

5. If all the brake shoe arms going to the spindles do not release, (refer to Figure 28.1), check the main cam gear and sliding arm alignment.

6. If the tape guide arm doesn't pull the videotape over to the capstan shaft (refer to Figure 15.10), check the pinch roller alignment.

7. If the videotape gets stuck in the unit or there is a tape already stuck in the unit and the roller guides are part way retracted, try to manually remove the tape.

8. If the gears won't move, remove the cassette carriage with the videotape in it. Then, start your alignment with the main cam gear and follow the directions from there.

9. Be sure to check for missing or bent teeth on each gear you remove for realignment.

Before removing any gearing for alignment, refer to the section in this chapter, "Checking the alignment."

Pinch-roller Alignment

This section covers models that use a cam gear in the lower portion of the pinch-roller assembly. For models that do not have a pinch-roller cam gear, refer to the section in this chapter, "Cam gear alignment." You might need to remove the cassette carriage in some units. Remove the top portion of the pinch-roller assembly (refer to Chapter 7, "Removing pinch rollers," and see Figures 7.14 and 7.15). This section will show how to align the gears in pinch-roller assemblies.

First Type

In this model:

1. Remove the cassette carriage (refer to Chapter 8 on "Removing and servicing cassette carriages"). The tape-guide assembly (refer to Figure 15.2) pulls the videotape up to the capstan shaft so that the pinch roller can clamp the videotape up against the capstan shaft.

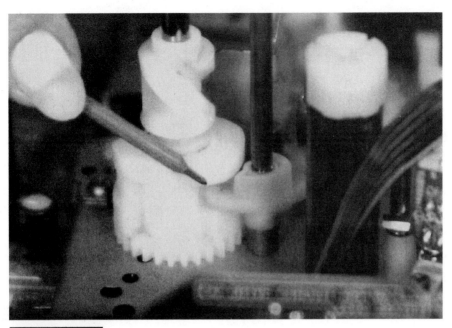

FIGURE 15.1

Tracking arm in the groove of the pinch roller cam gear.

2. Before removing any parts, notice how the tracking arm of the tape-guide assembly fits into the groove of the pinch roller cam gear, as shown in Figure 15.1.

3. Remove the O-ring on top of the tape-guide assembly, as shown in Figure 15.2.

4. Pull the back half of the tape-guide assembly (the white piece), along with the cam gear, up and off the two shafts.

5. Plug the unit in and turn the VCR on.

6. Place your fingers over the bubble on the take-up tape sensor, blocking all light to the tape sensor, as shown in Figure 15.3. This will cause the loading motor to run and reset the cam gear in the undercarriage.

7. Hold your finger over the tape sensor until the drive belt or worm gear stops turning, but the motor is still running.

8. Let go of the tape sensor and the motor will stop.

9. Unplug the unit.

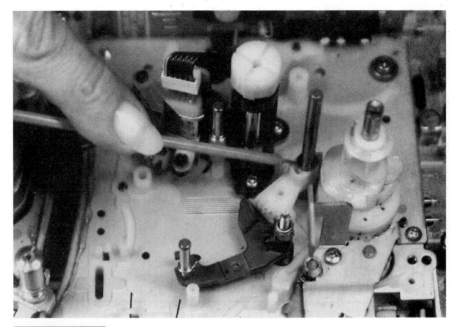

FIGURE 15.2

O-ring on top of tape guide arm assembly.

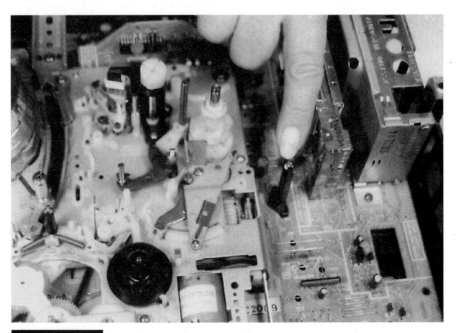

FIGURE 15.3

Take-up tape sensor.

10. Place the tracking arm of the tape-guide assembly into the groove of the cam gear and slide the two pieces down the shafts (refer to Figure 15.1).

11. You can move the two pieces up and down the shafts to align the gears.

12. Align the dot or little hole of the upper half of the tape-guide assembly to the first tooth on the lower half of the tape-guide assembly, as shown in Figure 15.4.

13. Align the first dot on the base of the cam gear to the notch on the bracket on the transport that is above the drive gear, as shown in Figure 15.5.

14. Push the cam gear down to align the teeth on the two gears.

15. Remount the O-ring on top of the tape-guide assembly.

16. Remount the top portion of the pinch roller assembly.

17. Before remounting the cassette carriage, be sure that the video tape holder is in the Up position.

Refer to the section in Chapter 8, "Remounting cassette carriages."

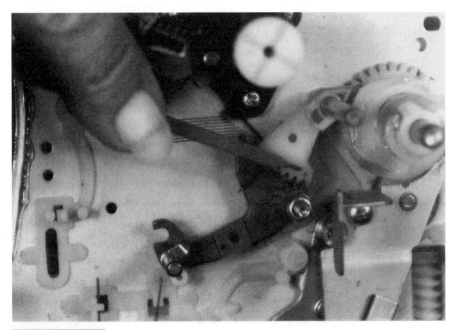

FIGURE 15.4

Align the dot to the first tooth on the tape guide arm assembly.

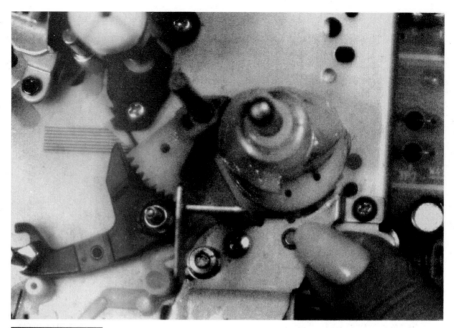

Align the first dot on the cam gear to the notch on transport.

Second Type

In this model:

1. Remove the loading-motor housing bracket that covers the gearing by removing the three screws in the bracket, as shown in Figure 15.6 (I drew arrows pointing at the mounting screws).

2. Lift the housing bracket straight up and the bracket and the loading motor will come right out.

3. Flip the housing bracket over and lay it on top of the transport. Wires are attached to the bracket or you can unplug the wires attached to the bracket.

4. Inside the housing bracket is the loading motor, a worm gear, and worm wheel, as shown in Figure 15.7. These gears need no alignment (refer to the section in this chapter, "Gear alignment").

5. Looking at the transport, you will see another wheel gear that needs no alignment, as shown in Figure 15.8. A wheel gear has no markings or little holes in it for alignment. You need to lift the wheel gear up and off its shaft because the teeth on the wheel gear cover the teeth on the cam gear.

FIGURE 15.6

Mounting screws to the loading motor housing bracket.

FIGURE 15.7

Loading motor, worm gear, and worm wheel.

FIGURE 15.8

Wheel gear.

6. Remove the pinch-roller cam gear by pulling the gear straight up and off its shaft.

7. Check the tape guide cam gear for alignment. Look for a little hole on top of this gear, close to the teeth. If it's aligned right, you will see another hole the same size in the transport. These two holes need to be right on top of each other, as shown in Figure 15.9. If the cam gear isn't aligned properly, then you will need to check or align the main cam gear. Refer to the section in this chapter, "Main cam gear alignment."

8. If the main cam gear is aligned and the tape guide cam gear still isn't aligned properly, remove the tape-guide arm assembly by pulling the tape guide arm straight up and off its shaft (see Figure 15.10). Be careful, a spring is at the base of the tape-guide arm.

9. Lift the tape-guide cam gear up and turn it until the holes align.

10. Push the cam gear back down to slide the teeth into the drive gear on the transport.

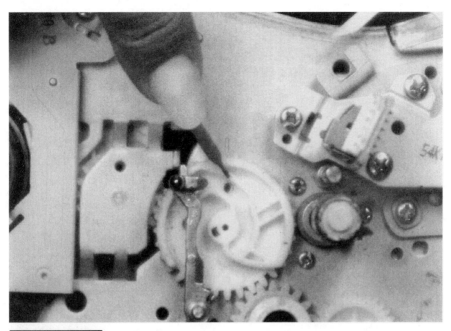

FIGURE 15.9

Aligning the hole in the tape guide cam gear to the hole in the transport.

FIGURE 15.10

Tape guide arm and tension spring.

Aligning the dot on the pinch roller cam gear to the arrow on the tape guide cam gear.

11. Slide the tape-guide arm back down its shaft and mount the spring at the base of the arm.

12. Align the pinch-roller cam gear.

13. On the face of the cam gear, right above the teeth, is a small hole or dot. Slide the pinch-roller cam gear down the shaft and align the dot to the arrow on the tape-guide cam gear, as shown in Figure 15.11.

14. Remount the wheel gear to the transport (see Figure 15.8).

15. Remount the housing bracket by placing the bracket over the pinch-roller cam gear and aligning the screw holes.

16. Push the housing straight down.

17. The plug at the back of the housing will plug directly into the mother board behind the transport.

Third Type

In this model:

1. Remove the lower pinch-roller bracket covering the pinch-roller cam gears.

2. Remove the screw holding the bracket down, as shown in Figure 15.12.

3. Push in on the clip in front of the bracket; at the same time, lift the bracket straight up and out, as shown in Figure 15.13.

4. This unit has two pinch-roller cam gears. The smaller cam gear moves the pinch roller back and forth. The larger cam gear moves the pinch roller up and down. The base of the large cam gear holds the smaller cam gear down. Release the latch at the base of the large cam gear, as shown in Figure 15.14. Pull back on the latch and lift the gear straight up and off its shaft.

5. Check to see if the smaller cam gear is aligned.

6. The top of the gear has an arrow-type hole. This arrow on the gear needs to be exactly aligned to the arrow on the circuit board under it, as shown in Figure 15.15. If not, then you will need to check or align the main cam gear. Refer to the section in this chapter, "Main cam gear alignment."

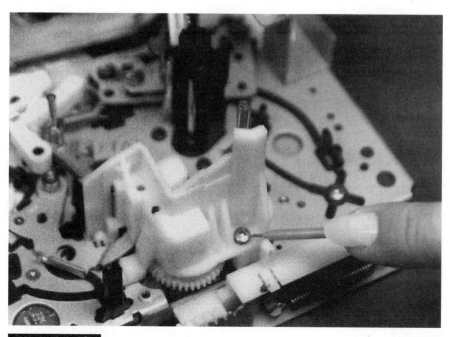

FIGURE 15.12

Mounting screw to the lower pinch roller bracket.

FIGURE 15.13

The clip holding down the lower pinch roller bracket.

FIGURE 15.14

Latch holding down the pinch roller cam gears.

FIGURE 15.15

The arrow on the cam gear pointing at the arrow on the circuit board.

7. After aligning the main cam gear, if the smaller cam gear is still out of alignment, lift the gear up and turn it to align the arrows.

8. Slide the gear back down its shaft.

9. The larger cam gear has a hole on top of the base. Slide the larger cam gear down the shaft and align the hole at the base to the notch on the smaller cam gear, as shown in Figure 15.16.

10. After aligning the two gears, push the large cam gear the rest of the way down until it latches.

11. Remount the lower pinch-roller bracket by aligning the pins and clip on the bottom of the bracket to the holes in the transport.

12. Push down on the bracket until the clip snaps into place.

13. You might need the aid of a flathead screwdriver to push on top of the clip, for it to snap into place.

14. Remount the mounting screw.

FIGURE 15.16

The alignment hole on the base of the large cam gear pointing at the notch on the smaller cam gear.

Loading Gear Alignment

This section deals with all the gearing that loads the video tapes to the video tape path. Refer to the section in this chapter, "Removing gears, sliding arms, and brackets." Start with the roller guides. Look at the position of the roller guides. They should be retracted all the way (refer to Figure 12.3). Next, look at the undercarriage at the base of the roller guides, as shown in Figure 15.17. You will see two hinge levers attached to each roller guide base. Each set of levers is attached to a loading gear. The two loading gears are attached together. To align these two gears to each other, look for a small hole in each gear close to the teeth. Position the gears so that the two little holes are facing each other, as shown in Figure 15.18. These marks can also be lines, arrows, dots, or a combination, such as an arrow pointing at a dot or hole.

A sliding gear or arm is attached to one of the loading gears. In this case, an arrow is on the arm and a very tiny arrow is on the gear. Align the two arrows so that they are pointing at each other, as shown in

FIGURE 15.17

The bottom of the roller guides and the four hinged loading levers.

FIGURE 15.18

Aligning up the two holes on the gears.

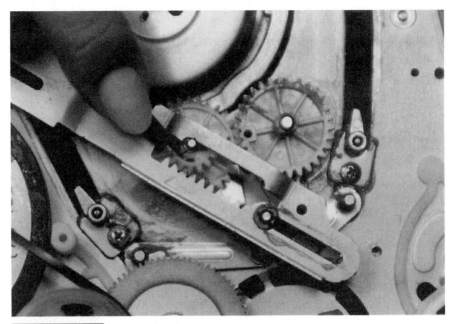

FIGURE 15.19

The arrow on the sliding arm pointing at the tiny arrow on the gear.

Figure 15.19. Other units could have a line on the gear and a line on the arm facing directly at each other, as shown in Figure 15.20. Some units could have a dot on the gear and a notch on the sliding gear facing directly at each other, as shown in Figure 15.21.

At the other end of the sliding gear or arm is a pin that fits down into the groove of the main cam gear, as shown in Figure 15.22. On the other hand, a hinged lever could be attached to the sliding gear and the hinged lever has a pin under it that fits down into the groove of the cam gear, as shown in Figure 15.23. In a few models, a gear is attached at the other end of the sliding gear. Refer to the section in this chapter, "Sliding arm alignment."

Main Cam Gear Alignment

Most moving parts are connected directly or indirectly to the main cam gear. If you aren't sure where to start the alignment, start it here. In most models, the loading motor drives a worm gear that drives the main cam gear. A few models have some gearing between the main cam gear and the worm gear. If this is the case, start your alignment af-

FIGURE 15.20

Aligning up the line on the sliding arm to the line on the gear.

FIGURE 15.21

Aligning up the notch on the sliding arm to the dot on the gear.

FIGURE 15.22

The pin under the sliding arm is attached to the groove of the cam gear.

FIGURE 15.23

The point where the pin under the hinged lever is attached to the groove in the cam gear.

ter the worm gear or worm wheel. In most cases, you will need to re-move a circuit board or a cam gear-mounting bracket covering the cam gear before you can align the gear. Refer to the sections in this chapter, "Removing a circuit board mounted to the transport" and "Removing gears, sliding arms, and brackets." Before removing any gearing, refer to the section in this chapter, "Checking the alignment."

The main cam gear is the heart of the gear system. The cam gear can make different gears move in opposite directions at the same time. This is performed by sliding arms attached to the grooves in the gear (refer to Figure 15.22). Some cam gears have grooves on both sides. When remounting the cam gear, be sure that all the sliding arm pins are in their groove.

In some models, the cam gear is mounted on top of the transport under the loading motor. Refer to the section in Chapter 27, "Third type" under the subhead "Removal of pinch-roller brackets." In the undercarriage, under the cam gear, there will be a slot on the transport and a pin that protrudes out from the top of a sliding arm. This pin protrudes through the slot and into the groove of the cam gear located on top of the transport, as shown in Figure 15.24. On the other hand,

FIGURE 15.24

The slot in the transport where the pin of the sliding arm protrudes into the groove of the cam gear.

FIGURE 15.25

The slot in the transport where the pin of the pinch bracket lever protrudes into the groove of the cam gear.

the cam gear can be in the undercarriage. Under the loading motor housing bracket is a slot in the transport and the pinch-roller bracket assembly has a pin protruding out of the bottom of the bracket. The pin protrudes through the slot and into the groove in the cam gear located in the undercarriage, as shown in Figure 15.25. I have removed the loading motor housing bracket on top of the transport so that you can see the cam gear and slot in the transport.

To align the cam gear, start by looking for a small hole near the edge of the gear, as shown in Figure 15.26. Next, look for another hole exactly the same size in the transport under the cam gear, a bracket on top of the gear, or another gear under or on top of the cam gear. In this case, a hole is in the sliding arm or gear above the cam gear. Align the two holes on top of each other so that you can see straight through both holes, as shown in Figure 15.27. In most models, the hole in the cam gear aligns up with the same size hole in the transport, as shown in Figure 15.28. In some models, a hole is on the edge of the cam gear and it aligns with the same size hole in the bracket above the cam gear, as shown in Figure 15.29.

FIGURE 15.26

A small alignment hole near the edge of the cam gear.

FIGURE 15.27

Looking straight through the hole on the sliding gear and the hole in the cam gear.

FIGURE 15.28

Looking straight through the hole in the cam gear and the hole in the transport.

FIGURE 15.29

Looking straight through the hole in the bracket and the hole in the cam gear.

FIGURE 15.30

Pointing at the large prong of the three-prong hole in the
center of the cam gear.

In a few models, in the center of the cam gear, a three-prong shaft
protrudes up from the undercarriage to the cam gear on top of the
transport. One of the prongs is larger than the other two. The cam gear
has the same pattern, as shown in Figure 15.30. Align the patterns and
push the cam gear down the shaft. Be sure that the pin from the sliding arm is in the groove under the cam gear.

Gears attached to the cam gear. One side has a sliding gear with an
arrow on it. The arrow is pointing at a large dot near the teeth on the
cam gear, as shown in Figure 15.31. On the opposite side is another
gear with a very small hole in it that faces directly at the notch on the
cam gear (refer to Figure 15.36). If you accidentally align the wrong
alignment mark on the cam gear to one of the connecting gears, the
other gear will not have an alignment mark to align with. That's how
you can know that the cam gear is in wrong.

If there is a gear partially under the cam gear it will be connected to
the cam gear. You will either find a hole in both gears that will align
over the top of each other or an arrow on the transport and a dot on the
gear facing each other (see Figure 15.31).

FIGURE 15.31

Arrow on the sliding gear pointing at dot on the cam gear.

Two cam gears align with each other. Each cam gear has a hole in the gear and aligns with the exact same size hole in the transport. Plus, one of the cam gears has an arrow pointing right at the hole in the other gear, as shown in Figure 15.32.

Sliding Arm Alignment

Most sliding gears or arms are attached to a cam gear. The other end of the arm is attached to a small gear or to a hinged lever. The sliding arm can be under the cam gear or on top of the cam gear. Also a circuit board or a bracket could cover the sliding arm. Refer to the section in this chapter, "Removing a circuit board mounted to the transport" and "Removing gears, sliding arms, and brackets." Also read the last two sections in "Loading gear alignment."

In this model:

1. Under the cam gear is a large sliding arm, as shown in Figure 15.33. Remove the cam gear and slide the arm so that the roller guides are fully retracted.

FIGURE 15.32

The hole in each cam gear aligning up with the holes in the transport and an arrow pointing at the opposite cam gear alignment mark.

FIGURE 15.33

The alignment hole in the sliding arm and the pin that goes into the cam gear groove.

2. Look for the alignment mark.

3. A hole is at one end of the sliding arm. Align the hole in the arm with the same size of hole in the transport (Figure 15.33). On top of this arm, a pin protrudes up and goes into the groove on the bottom side of the cam gear.

In other models:

1. Remove the bracket with the cam gear mounted to it (refer to Figure 21.24).

2. Slide the sliding arm all the way to the end, where the roller guides are retracted.

3. Look for an alignment mark that will almost be aligned. The holes in the arm have no holes in the transport to align with, so there must be another alignment mark. In this case, the alignment mark is the line on top of the sliding arm.

4. Slide the arm over to align the thin line to the center of the shaft, as shown in Figure 15.34.

FIGURE 15.34

Aligning the thin line to the center of the shaft.

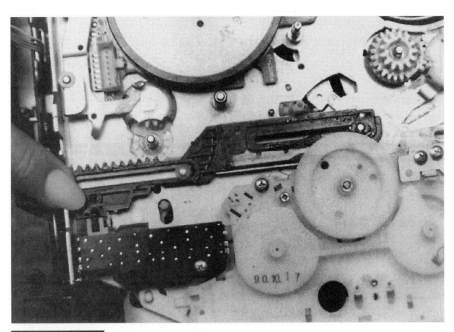

FIGURE 15.35

Aligning the hole on the sliding arm to the hole in the transport.

In some models:

1. Under the cam gear is a sliding arm and the arm is not connected to the roller guides. Remove the cam gear (refer to Figure 15.28).

2. Sliding the arm back and forth won't make the roller guides move. Instead, the arm is connected to other levers and the mode switch. The alignment mark is a hole in the arm over the same size hole in the transport, as shown in Figure 15.35.

In some models, a gear is attached to both ends of a sliding arm. The alignment mark is a small, thin line on the gear and a very small thin line right behind the teeth on the sliding arm. These lines directly face each other (refer to Figure 15.20). The other end has the same type of alignment.

Gear Alignment

If no marking is just behind the teeth on the gear, such as a hole, arrow, dot, post, or notch, the gear doesn't need to be aligned. Look very carefully at the gear because the mark can be quite small (refer to Figure 15.37). Most worm wheels do not need to be aligned (refer to Figure 15.7).

In models that have a gear that is connected to a gear in the undercarriage and goes through the transport and is connected to a gear on top of the transport, the alignment marks could be on the top of the gear or the bottom of the gear or both sides.

Some gears have one gear on each side of the main gear. On one side, a small hole is close to the teeth, facing directly at a notch in the gear beside it, as shown in Figure 15.36. On the opposite side, a small hole is close to the teeth, facing directly at a dot on top of the other gear. If you try to align the wrong holes to the wrong gear, the other gear will not have an alignment mark to align to because the holes are offset.

This model has a very tiny arrow on one gear and a very tiny post protruding up, just inside the teeth of the gears, as shown in Figure 15.37. I have removed the center gear so that you can have a better

FIGURE 15.36

Aligning the little hole to the notch on the big gear and on the other side, aligning the big hole to the dot on the little gear.

FIGURE 15.37

A tiny arrow and tiny post.

view. The center gear has a notch that aligns with the tiny arrow and the hole in the gear aligns with the post.

Cassette Carriage Alignment

To align the gear on the cassette carriage, the cassette holder has to be in the Up position. To remove a gear or bracket, refer to the section in this chapter, "Removing gears, sliding arms, and brackets." Gears are on each side of the carriage. Align the gears in the same way. Refer to the section in this chapter, "Gear alignment."

If you do not find any alignment marks on a gear and its teeth, don't go all the way around, then align the gear with the alignment mark to the first tooth of the gear with no mark. If no alignment mark is on the drive gear of the carriage, no alignment mark is on the gear on the transport, and

> **TRADE SECRET**
>
> The one exception is that a dot or an arrow could point at the first tooth, as shown in Figure 15.38.

FIGURE 15.38

An arrow pointing at the first tooth.

teeth run all the way around both gears, then be sure that the cassette holder is all the way forward and slide the two gears together. For more information, refer to the section in Chapter 8, "Remounting the cassette carriage."

Take-up Spindle Alignment

Some direct-drive units have a white gear beside the take-up spindle. The white gear is directly connected to a reel motor that drives the take-up spindle.

When you put this system in Fast Forward, the take-up spindle doesn't turn and you hear a loud grinding noise. In Play, you'll hear a clicking noise. In either case, after 60 seconds, the unit stops. The problem is that the gear has worked itself up or down the shaft of the reel motor (refer to Figure 19.17). The teeth on the gear come out of alignment with the teeth on the take-up spindle. To correct this problem, place a flathead screwdriver under the gear and pry the gear up the shaft until the teeth on the gears are realigned.

Checking Your Alignment

After you have finished the alignment and the transport is still removed from the unit, you need to check it. On the other hand, you can check the alignment before you align the gearing by running the transport through its cycle.

To make the gears move:

1. Be sure that the cassette carriage is removed from the transport. (Refer to the section in Chapter 8, "Cassette carriage removal.")

2. Locate the loading motor and spin the pulley or worm gear attached to the loading motor. Refer to the section in Chapter 19, "Loading motors."

3. Rotate the pulley or worm gear clockwise until the roller guides on top of the transport have moved into their V-mounts tightly and the pinch roller has moved over to the capstan shaft tightly.

4. Spin the pulley or worm gear in the opposite direction until the roller guides have completely retracted. If you cannot complete the cycle and the transport jams up before the roller guides are in their V-mounts or if the pinch roller isn't tight against the capstan shaft, a gear is in wrong or a bad gear has bent or missing teeth.

5. Rotate the pulley or worm gear in the opposite direction until the alignment mark on the main cam gear is aligned. Always start the alignment after the worm gear that is attached to the loading motor.

6. Check each gear to see if they are aligned properly.

If you are just starting out and the cam gear alignment mark is off by a little bit and the roller guides are retracted back all the way, then rotate the pulley or worm gear to align the cam gear. Then check all the rest of the gearing. If the alignment mark is way off and the roller guides are retracted, then remove the cam gear and start the alignment from there.

Removing a Circuit Board Mounted to the Transport

To remove a circuit board mounted to the transport:

1. Remove the belt off the flywheel (refer to Figure 10.2).

2. Remove any wires soldered to the circuit board and draw a map of where the wires are connected (refer to the sections in Chapter 17, "Using solder wick to remove old solder" and "Soldering pins to a circuit board"). Refer to the last section in Chapter 31, "Repairing cracks on circuit boards." Read the last paragraph.

3. Remove all screws mounted to the board.

4. Look on top of the circuit board for any rectangular or triangular holes.

Read the following and use the sections that pertain to your unit:

1. Two rectangular holes, as shown in Figure 15.39. Two connections are between these two holes. Unsolder the two connections. (A connection is where a lead from a component, such as a sensor, passes through a little hole in the circuit board and is soldered to the circuitry side of the board and a track is connected to the same connection.)

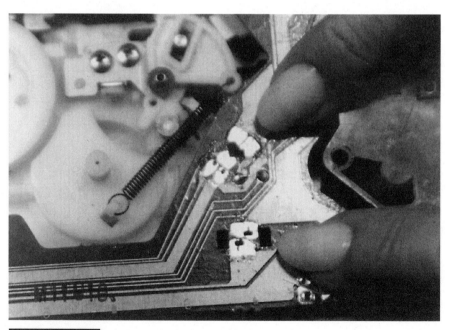

FIGURE 15.39

Two rectangular holes or one triangular hole showing the pins to unsolder.

2. One triangular hole has four or five soldering connections around the hole, as shown in Figure 15.39. Unsolder these connections.

3. Two rectangular holes have black small clips protruding up through the holes and fastened to the board with connections between them or around them (refer to Figure 21.10). This means that the part is fastened to the board and you do not need to unsolder the part.

4. Some units have a mode switch soldered to the circuit board. You need to unsolder the connections on the board (refer to Figure 21.15).

5. Look for any clips or latches that hold the board down. Refer to the section in Chapter 32, "Removing front covers." This section shows the different types of latches and how to release them.

6. Hold the board and try to pick it up. If any section of the board doesn't move, then you need to disconnect or unsolder something.

7. Recheck the board at that location.

8. Carefully pick up the loose part of the board and look under it where it is fastened down. A plug could be attached to the transport (same type of plug, as shown in Figure 31.18). When removing the board in some units, tape sensors might be mounted to the board (refer to Figure 15.3).

In other units:

1. Ribbons are attached to the board. Unplug the ribbons (refer to Figure 18.12). Lift the board straight up.

2. The backside of the board has one to three sets of wires attached to the board.

3. Unplug the plug attached at one end or the other on these sets of wires. Don't worry, each plug will be a different size.

4. When remounting, be sure to plug in all plugs going to the bottom of the board.

5. Be sure to align the pins on the transport, as shown in Figure 15.40, to the holes on the board and solder them.

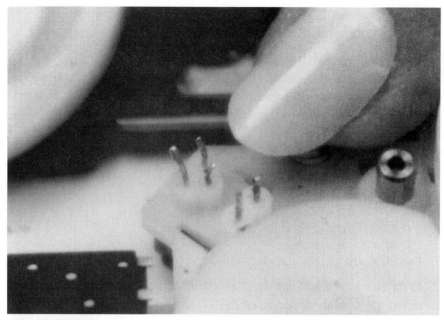

FIGURE 15.40

Pins on the transport to be soldered to the circuit board.

6. Be careful when sliding the tape sensors back through the holes in the transport.

7. Be sure to re-solder any wires back onto the board.

8. Remount the mounting screws.

Removing Gears, Sliding Arms, and Brackets

To remove a gear for realignment, remove the E- or O-ring from the shaft in the middle of the gear. Refer to the section in Chapter 7, "Removing E- and O-rings." Other gears are attached by a latch at the top of the gear at the shaft. Pull the latch out and lift the gear up and off the shaft (see Figure 15.41).

To remove sliding gears or arms, remove the E- or O-ring from each shaft protruding up through the sliding arm (refer to Figure 15.20). Usually, they have no more than two rings. Some models have a screw with a large washer on it that holds one end of the sliding arm down (refer to Figure 15.34). Remove the screw.

Other models have a bracket or gear that holds down the sliding arm. If it's a gear, remove the E- or O-ring and lift the gear off. If it's a

FIGURE 15.41

Prying back the latch on top of the gear.

bracket, remove the two or three mounting screws and lift the bracket off. Wires might be above the bracket. If so, tabs will hold the wires in place (refer to Figure 24.28). Just slide the wires out from under the tabs and move them to the side.

In other models, you will need to remove the loading motor and housing above the cam gear. Just remove the mounting screws in the housing, unplug the loading motor, and carefully lift the housing straight up and off. Refer to the section in Chapter 27, "Removal of pinch roller brackets" and refer to the section in Chapter 19, "Removing loading motors and belts."

In a few models, you will need to remove the flywheel. Refer to the section in Chapter 25, "Removing the capstan shaft in direct drive VCRs."

Review

1. The marking for a gear is just behind the teeth. Look for a hole, arrow, dot, post, or notch.
2. Each alignment mark will be the only mark of that type on the gear.

3. To remove a gear for realignment, remove the E- or O-ring from the shaft in the middle of the gear.

4. Position two gears so that the two alignment marks are facing directly at each other.

5. To remove sliding gears or arms, remove the E- or O-ring from each shaft, which protrudes up through the sliding arm.

6. A sliding gear or arm is attached to a gear. Align the alignment marks so that they are pointing at each other.

7. Align the hole in the cam gear to a hole above in a bracket or a hole above in another gear or to the hole below in the transport.

8. Gears attached to the main cam gear. Position two gears so that the two alignment marks directly face each other.

9. When remounting the cam gear, be sure that all the sliding arm pins are in their groove.

10. To remove a bracket, remove the two or three mounting screws and lift the bracket off.

11. When removing a circuit board, look for any white, thin lines that make a cross with one triangular or two rectangular holes, which show that the pins need to be unsoldered.

12. When remounting a circuit board:
    A. Be sure to plug in all plugs.
    B. Align the pins on the transport to the holes on the board and solder them.
    C. Be careful when sliding the tape sensors back through the holes in the transport.
    D. Resolder any wires back onto the board.
    E. Remount all mounting screws.

Replacing Fuses

A VCR **uses** anywhere from one to six fuses. The fuses snap into a holder and are easy to replace. Before proceeding with this lesson, unplug the unit. Be sure to read the section in Chapter 2, "Electrical shock."

Locating the Main Power Fuse

If the VCR goes completely dead, nothing will light, including the digital clock. If this situation occurs, check the main power fuse. First remove the top main cover. Locate the main fuse by following the A/C cord to where it enters into the back of the unit. From there, the A/C cord goes either into a metal or plastic covered box or onto a circuit board.

Older VCRs might have a clear plastic-covered box to hold the fuses. You can see the fuses inside the box, as shown in Figure 16.1. Simply unsnap the plastic cover and pull it to the side to remove the fuse.

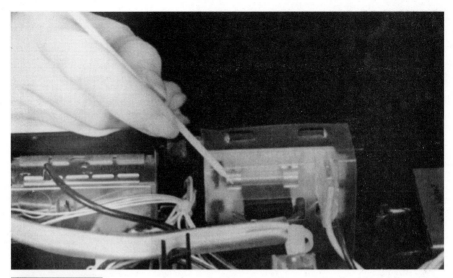

FIGURE 16.1

Locating the main power fuse in a transparent covered box.

Other models could have an opaque plastic box. In this case, follow the AC cord into the unit, where it enters the plastic box (see Figure 16.2).

Mounting screws might be on each side at the base of the plastic box. Pull the plastic cover straight up and off, as shown in Figure 16.2. You'll find a small circuit board under the plastic box with the fuse mounted to it, as shown in Figure 16.3.

In some newer models, the AC cord goes into a metal box. The entire power supply and fuse is located inside this metal box. You need to remove the power supply to replace the fuse. It has two or three mounting screws. One screw is located in front at the base of the power supply and the other is located on top at the rear of the power supply, as shown in Figure 16.4. One mounting screw runs up through the bottom plate under the VCR.

After removing the mounting screws, hold the power supply and pull it straight up and out, as shown in Figure 16.5.

When remounting the power supply, align the two plastic pins on the chassis to the pinholes on top of the power supply, as shown in Figure 16.4. The plug under the power supply will automatically line up. Just

FIGURE 16.2

Pulling off the plastic cover.

push down on the power supply until the screw holes on the chassis are up against the mounting holes on the power supply. Replace the mounting screws.

In older models that don't use a covered box, the fuse is directly mounted to the power-supply circuit board. Follow the AC cord to where it connects to the circuit board. Right beside the connection is a snap-in fuse. This fuse is the main power fuse. If you find more than one fuse mounted to the circuit board, the main power fuse

> **TRADE SECRET**
>
> A plug is under the power supply that plugs into the motherboard. This plug automatically unplugs when lifting up the power supply. The location of the fuse is shown in Figure 16.6.

FIGURE 16.3

Locating the main power fuse on a circuit board.

is closest to the AC cord connection, as shown in Figure 16.7. The other fuses are secondary fuses.

Newer models only have one fuse, which is located beside the AC cord, where it is connected to the motherboard, as shown in Figure 16.8.

QUICK REPAIR

To remove a snap-in fuse, place a small flathead screwdriver under one end of the fuse and pry the fuse up, using a circuit board as a lever, as shown in Figure 16.9.

Checking a Fuse

A fuse can go bad in one of two different ways. If a fuse blows out, it turns black inside the glass tube, as shown in Figure 16.10. This condition usually means that the VCR has a dead short. If you replace the fuse, it will just blow out again. If

FIGURE 16.4

Mounting screws and alignment pins to the power supply.

FIGURE 16.5

Removing the power supply.

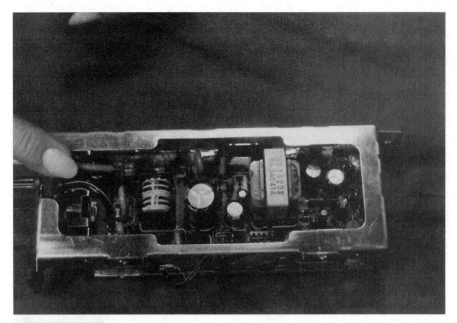

Locating the main power fuse inside the power supply.

Locating the main power fuse.

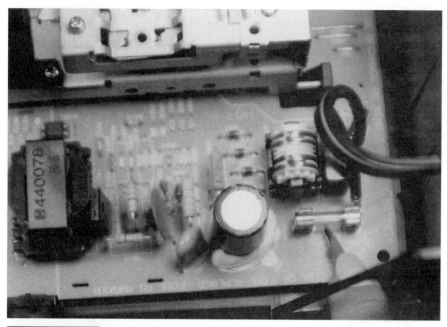

FIGURE 16.8

Locating the main power fuse on the motherboard.

FIGURE 16.9

Prying out the fuse.

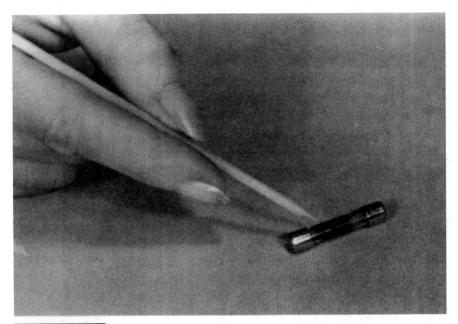

FIGURE 16.10

A bad fuse.

this is the case, you should take the unit to your local service center for circuit repairs.

The second way is that the fuse opens, which means that the thin wire inside the glass tube has broken. Sometimes it's hard to tell if the wire is broken. To test for a broken wire, remove the fuse and tap the glass tube. If the wire inside starts vibrating, the fuse is bad.

A power surge to the unit will usually cause an open fuse. If the fuse has opened, you can go ahead and replace it.

Some fuses have a white coating inside of the glass. It's not possible to check inside this type of fuse to see if it's blown. You can check the continuity of the fuse with an ohmmeter or take the fuse to your local service center and have it checked.

TRADE SECRET It's possible for the fuse to look good from the outside, but have a bad connection on the inside.

Secondary Fuses

In older models, the secondary fuses are located beside the main power fuse. You might find one to five more fuses (refer to Figure 16.7). These fuses are on the component side of the power-supply circuit board. These fuses run on a low voltage; each fuse supplies a separate portion of the unit. If a certain function stops working, but the VCR light or only part of the unit functions properly, check the secondary fuses. If you find a blown fuse, the VCR has a circuit problem. If the fuse has just opened, replacing the fuse will probably fix the problem.

In newer models, the secondary fuses are called *IC protectors*; they look like a little black transistor, except that they have two leads instead of three, as shown in Figure 16.11. These types of fuses are soldered to the circuit board and will have to be unsoldered to be removed. A unit might have four or more IC protectors. In most cases, you will not be able to locate these types of fuses without a service manual. If you do locate any IC protectors, you will need to check the continuity by placing the ohmmeter probes across both leads while the IC protector is still in the unit. The ohmmeter should read near zero ohms if the fuse is good or read 1000 ohms and up if it is bad.

Reading the Value of a Fuse

To read the value of a fuse, look for printing on each metallic end of the fuse. For example, one end of the fuse might read *125V5A*. This label means the fuse has a voltage of 125 volts and amperage of 5 amps. Voltage is the measure of electrical pressure. Amperage is the strength

FIGURE 16.11

An IC protector.

of the electrical current. Read all glass-tube fuses in the same manner. To replace a bad fuse, refer to the section in Chapter 1, "Parts," or go to your local service center.

Review

1. Always unplug the unit before replacing a fuse.

2. Most VCRs have one fuse and some have as many as six fuses inside.

3. To find the main power fuse, always follow the AC cord to the covered box or to a circuit board.

4. A blown fuse turns black and an open fuse just breaks the wire inside of the glass.

5. All secondary fuses are located on a circuit board close to the main power fuse.

6. An IC protector is soldered in and can be located anywhere in the unit.

Replacing the Audio Head

Connect the VCR to the TV monitor and tune in a local broadcast channel on the VCR (refer to the section in Chapter 5, "Connecting the VCR to a TV monitor"). Check the audio portion of the program. If the audio portion is functioning properly, then insert a videocassette and push Play. If the picture is clear but the audio is malfunctioning, you have to determine if the A/C head is out of alignment, the audio head is bad, or the audio circuit is not working. Remember that the audio head is incorporated in the A/C head.

Checking for a Bad Audio Head

Push Play and turn up the volume control on the TV monitor. If you have a stereo hi-fi VCR, put the stereo switch into the Normal position. Using the flat surface of a small flathead screwdriver, gently tap on the face of the A/C head while the videotape is playing, as shown in Figure 17.1. Tap on the top right corner of the A/C head. You should hear a popping sound coming from the speaker in the TV monitor. For reference on the audio portion of the A/C head, refer to Figure 13.1.

FIGURE 17.1

Tapping on the top right hand side of the A/C head.

If you hear a popping sound, the videotape isn't making a flush contact with the A/C head. This condition occurs for two reasons. The first reason is a buildup of residue on the A/C head. In this case, just clean the A/C head. Refer to the sections in Chapter 4, "A/C head and eraser head" under the subhead "Cleaning the various components." The second reason is that the face of the A/C head might have a small dent from being improperly cleaned. The surface of the A/C head dents easily. Don't use a hard or rough object to clean off the residue buildup. The dents cause the videotape to make improper contact with the audio portion of the A/C head. If an A/C head has been dented, it will have to be replaced.

If you do not hear a popping sound, go to the small circuit board connected to the back of the A/C head. Place your finger on the shaft of a screwdriver and touch the tip of a screwdriver to each contact (soldered connection) directly behind the audio portion of the A/C head. The two top right pins that protrude through the circuit board behind the A/C head are the contacts to touch. Touch each contact,

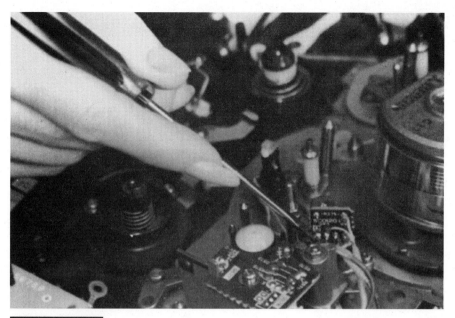

FIGURE 17.2

Touching each contact on the back of the A/C head.

one at a time, as shown in Figure 17.2. Do not short any of the contacts together.

You should hear a loud buzz from the speaker when touching the correct contact. You must have your finger on the screwdriver to create the buzzing sound. If you do hear a buzz, the audio circuit is working. One or two contacts will cause this buzz. The buzz should be louder than the normal volume. If the A/C head is buzzing, but not popping, the A/C head is bad. If it's not buzzing, the audio circuit has malfunctioned and you will have to take the unit into your local service center for circuit repairs.

To determine if the A/C head is out of alignment, use the tapping method on the face of the A/C head:

1. Listen for the popping sound. This sound shows you that the A/C head is functioning.

2. Next, check the setting of the A/C head. The ideal setting for an A/C head is straight up and down, not tilted.

3. The top edge of the videotape should barely touch the centerline of the audio and erase head (see Figure 13.2).

If the setting of the A/C head looks like it's not right, refer to Chapter 13, "Audio head alignment."

Removing an Audio Head

Unplug the unit before proceeding with this section. Older models have either a white or brown plug located on the top of the little circuit board mounted to the back of the A/C head, as shown in Figure 17.3. Remove this plug. In newer models, a ribbon is attached to the top of the circuit board. Pull up on the locking tab where the ribbon connects, then pull the ribbon out of the connection (refer to Figure 9.17). The end of the ribbon will pull out of the plug.

All A/C heads are mounted on a bracket with three adjustment screws. In newer models, all three adjustment screws are springloaded and the springs are located between the upper and the lower brackets (refer to Figures 13.8 and 13.9). If you have this type of unit, proceed to the section in this chapter, "Replacing an audio head with a spring-loaded mount." In other models, no springs are between the

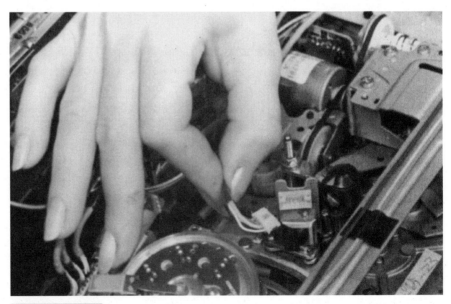

FIGURE 17.3

Removing the plug from the A/C head.

upper and lower mounts (refer to Figure 13.6). Remove the Phillips head spring-loaded screw (the one with the spring wrapped around it) located on the right side of the A/C head (refer to Figure 13.10). Remove the azimuth adjustment that is the Phillips head screw on the left side of the A/C head (refer to Figure 13.6). Do not remove or turn the tilt-adjustment screw or Allen nut directly behind the A/C head (Figure 17.10). This adjustment is crucial and is used for remounting the A/C head. After removing the two screws, lift the A/C head straight up. The mounting bracket will come up with it. Turn it around and look at the circuit board mounted on the back of the A/C head.

I've taken an A/C head out of another unit so that you can see what the back of an A/C head looks like without a circuit board, as shown in Figure 17.4. Notice that six pins protrude out the back of the A/C head. These pins go through a small circuit board and are soldered to it. To remove the circuit board from the A/C head, remove the solder from each pin. In stereo models, the A/C head has 10 pins instead of six—all protruding out the back.

FIGURE 17.4

Pins protruding out the back of an A/C head.

FIGURE 17.5

Using solder wick to remove old solder.

Using a Solder Wick to Remove Old Solder

You'll need a soldering iron or soldering gun and some solder wick. Look at the circuit board where the pins come through.

1. Place the solder wick over one of the soldered pins.

2. Place a hot soldering iron on top of the solder wick, as shown in Figure 17.5. As the wick gets hot, it will absorb the old solder.

3. Remove the iron and move the solder wick forward to a clean spot on the solder wick.

4. Place a hot iron back onto the wick to soak up the remaining old solder. You might have to repeat this process two or three times to remove all of the solder. Do each pin the same way.

Some models have a wire attached to the same connection as the pin. In this case:

1. Touch the hot soldering iron to the connection and pull the wire off, as shown in Figure 17.6.

FIGURE 17.6

Removing a wire from a circuit board.

2. Use solder wick to remove the old solder.

3. It is helpful to draw a diagram. The diagram should show the color of the wire and to which pin it was connected.

No more than one wire and one pin should be soldered to any one connection.

After removing the solder from each pin, simply pull the A/C head off the circuit board, as shown in Figure 17.7. Refer to the section in Chapter 1, "Parts." Then, proceed to the next section. Do not lose the old A/C head; you will need it later.

> **QUICK»TIP**
>
> You usually shouldn't have to remove more than two wires from the back of the circuit board. All other wires don't have to be removed; they're connected to the circuit board.

Soldering Pins to a Circuit Board

1. Take a new A/C head and push the protruding pins through the pinholes on the circuit board.

2. Coat the tip of a hot soldering iron with solder, as shown in Figure 17.8.

FIGURE 17.7

Removing the A/C head from the circuit board.

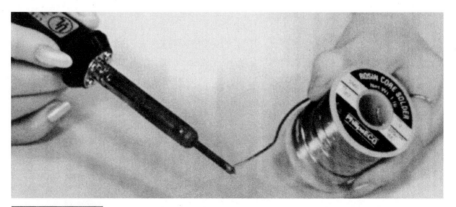

FIGURE 17.8

Coating the tip of the iron with solder.

3. After coating the tip, tap the excess solder off the iron onto a block of wood. Don't tap the excess solder over the VCR.

4. Position the tip of the iron to touch the side of the pin and the circuit board simultaneously.

5. Push the solder into this connection until it flows all the way around the pin, as shown in Figure 17.9.

6. Remove the solder and leave the iron on the connection for another two or three seconds to bond the solder to the circuit board and the pin.

7. Remove the iron and repeat this procedure for each pin.

QUICK REPAIR

To replace the wire to the pin connection, wrap the exposed end of the wire around the pin first. Then solder it as previously shown.

QUICK》》TIP

Don't solder two pin connections together. They must be separate. If you put too much solder on a pin, the solder might flow over and connect to a neighboring connection, creating a shorted circuit. Simply use the solder wick to absorb any excess solder.

FIGURE 17.9

Soldering a connection.

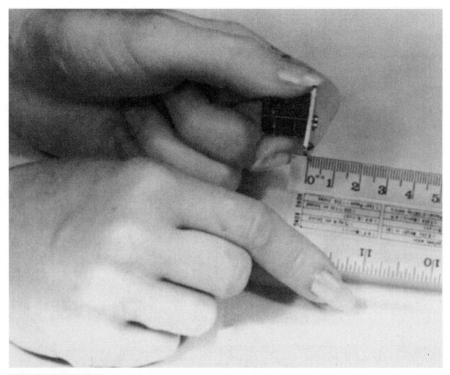

FIGURE 17.10

Measuring the tilt screw adjustment.

Remounting the Audio Head

Locate the tilt-adjustment screw or Allen nut that you left mounted in the old A/C head bracket. There are two ways to measure this screw. First, turn the screw counterclockwise to remove the screw and very carefully count every turn you make. Then, remounting the screw in the new bracket, turn the screw clockwise the exact same amount of turns and stop. The second way is to measure the distance that the tilt screw protrudes through the bottom of the bracket, as shown in Figure 17.10. This screw varies in length, shape, and type. Remove the screw from the old bracket and place it into the new bracket. Adjust this screw to the same measured distance as the old bracket. This step will give you a good starting point for realigning the A/C head.

1. Place the A/C head in its original position.

2. Place the spring on the proper setscrew before remounting the screw to the right side of the A/C head, as shown in Figure 17.11.

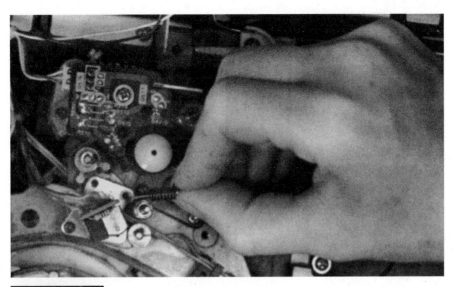

FIGURE 17.11

A spring-loaded screw.

3. Remount the azimuth screw on the left side of the A/C head.

4. Adjust these two set screws so that the A/C head sits as level as possible.

5. Be sure to plug in the plug on top of the A/C head.

Now you're ready to do an audio head alignment. Refer to Chapter 13, "Audio head alignment."

Replacing an Audio Head with a Spring-loaded Mount

Spring-loaded mounts use three spring-loaded adjustment screws, instead of one (refer to Figure 13.8). These springs are located between the upper and lower brackets under the A/C head.

1. For a reference point, measure the distance from the transport to the top of the upper bracket (this is the bracket the A/C head is mounted to), as shown in Figure 17.12. If the lower bracket sticks out further than the upper bracket, then measure the distance from the lower bracket to the upper bracket. Write this measurement down.

2. Remove all three screws and lift the springs and the A/C head straight up and out.

FIGURE 17.12

Measuring the distance between the A/C head mounting bracket and the transport.

3. Turn the A/C head around and remove the small circuit board.

4. Refer to the sections in this chapter, "Using a solder wick to remove old solder" and "Soldering pins to a circuit board."

When you install the new A/C head:

1. Put the spring over each hole of the lower bracket.

2. Place the upper bracket on top of the springs, align the holes in the upper bracket to the lower bracket, and remount each adjustment screw.

3. Measure the distance from the transport to the top of the upper bracket or from the lower bracket to the upper bracket as before. If the distance has altered, readjust all three screws. The measurement should be exactly the same as before and the A/C head should be sitting level.

4. Re-solder the pins and plug-in the plug or ribbon on top of the A/C head.

Now, you're ready to do the audio head alignment. Refer to Chapter 13, "Audio head alignment."

Review

1. Place the unit into the Play mode with videotape inserted.

2. Tap lightly on the audio portion of the A/C head to see if the speaker makes a popping sound. If so, check the face of the head.

3. If it doesn't pop, touch the audio connections on the back of the A/C head. Listen for a buzzing sound coming from the speaker.

4. If you hear buzzing, but no popping, the A/C head is bad and needs to be replaced.

5. To remove the A/C head, remove the two screws on each side of the A/C head, leaving the tilt-adjustment screw.

6. If you have three spring-loaded screws, measure the distance from the transport to the top of the old A/C head bracket before removing the bad A/C head.

7. Remove the old solder from the circuit board with the aid of a solder wick.

8. Place the pins of the new A/C head through the circuit board and solder them in.

9. Be sure that you haven't soldered two connections together.

10. Measure the distance of the tilt-adjustment screw on the old A/C head bracket and place that setscrew into the new bracket at the same distance.

11. Remount both screws on each side of the A/C head and adjust these screws to make the A/C head as level as possible.

12. Refer to Chapter 13, "Audio head alignment."

Replacing the Video Heads

he video heads are mounted to the upper video drum and are preset by the factory. When looking at the top of the upper drum, you will see two, three (mostly), or sometimes four small holes on each side (refer to Figure 6.16). Under these holes, on the bottom of the upper drum, is where the heads are mounted (refer to Figure 5.2). Some units have video-head adjustment nuts in these holes, as shown in Figure 18.1. Do not touch these adjustments because they are factory set. In most models, to replace the heads, you need to replace the upper video drum. In some models that have a grounding spring on top of the video drum, you need to remove the spring. Refer to the section in Chapter 22, "First location," under the subhead "Locating the high-pitched sound," Figure 22.2.

Two-headed Machine

In older models, the upper video drum comes right off. This type of unit has two mounting screws on top of the upper drum (refer to Figure 18.3). When replacing the upper video drum, never grab the

FIGURE 18.1

The adjustment nuts to the video heads.

drum where the video heads protrude out (refer to Figure 5.1). Read the section in Chapter 5, "The video heads" and the section in Chapter 6, "Cautions regarding the video drum." To order a new upper video drum, refer to the section in Chapter 1, "Parts." Read the following sections and follow the instructions that pertain to your unit.

First Type

A few models have a small plate or cover on top of the upper video drum, as shown in Figure 18.2.

1. Remove the two screws and lift off the plate. Most units don't have this plate. In most models, you only need to remove one set of two mounting screws on top of the upper video drum (refer to Figure 6.16).

2. Look at the top of the video drum to see how the heads are attached to the video head transformer inside the drum. The video head transformer is the part that rotates between the upper and lower video drums so that the video signal can be passed through a moving part.

FIGURE 18.2

The top cover and its mounting screws.

3. In this type of unit, four wires come up through holes on the upper drum. These wires are attached to a little round circuit board in the middle of the upper drum, as shown in Figure 18.3. There are two yellow wires, one red wire, and one brown wire. Unsolder these four wires (refer to the section in Chapter 17, "Using a solder wick to remove old solder"). After unsoldering the wires, pull the upper drum straight up and off.

4. Before remounting the upper drum, read the section in this chapter, "Remounting the upper video drum." Be sure to position the upper drum so that the colored wires on top of the video drum can be soldered to the same-colored wires coming up from the video transformer. The wires coming up from the transformer are connected to the little circuit board, as shown in Figure 18.4.

5. If the new upper drum has some different colored wires, then position the upper drum so that the red wire on top of the video drum can be soldered to the connection with the red wire coming up from the video transformer. If the new upper drum has two green, two gray, or two black wires, instead of two yellow wires, solder

FIGURE 18.3

Four video head wires attached to the circuit board in the center of the upper drum.

FIGURE 18.4

Four transformer wires attached to a circuit board above the video transformer.

the appropriate color to the yellow wires coming up from the video transformer. If the last wire is blue or white in color, instead of brown, solder one of these colors to the brown wire coming up from the video transformer.

Second Type

In other models:

1. Remove the two mounting screws on top of the drum.

2. A larger round circuit board is on top of the upper drum. There are two soldered pins to a set and two sets of soldered pins on each side of the circuit board above each video head. Unsolder the video transformer pins to remove the upper drum.

3. The video transformer pins are the two sets of pins furthest away from each video head, as shown in Figure 18.5. Unsolder these two sets of pins. After unsoldering the pins, pull the upper drum straight up and off.

FIGURE 18.5

The two sets of soldered pins attached to the video transformer.

4. Before remounting the video drum, read the section in this chapter, "Remounting the upper video drum."

5. On top of the upper drum, where the circuit board is located, notice that half of the board is white and the other half is green (refer to Figure 18.5). When remounting, position the upper drum so that the white part of the circuit board is directly above the white part of the board on the video transformer (Figure 18.7 shows the coloring of the circuit board without the pins).

6. Be sure to align the pins from the video transformer up through the pinholes in the circuit board on the upper drum.

Third Type

1. Some models have four wires protruding out of two holes in the upper drum. These wires are soldered to two small pins on each side above the video heads, as shown in Figure 18.6. It has two black wires, one red wire, and one blue wire. Unsolder all four wires. After unsoldering the wires, pull the upper drum straight up and off.

2. Before remounting the video drum, read the section in this chapter, "Remounting the upper video drum." On top of the video drum are two colored posts with two small pins on top of the posts (refer to Figure 18.6).

3. When remounting, note that these posts are connected to the video heads. One post is red and the other is black. Two sets of wires are connected to each side of the transformer, as shown in Figure 18.7.

4. Position the upper drum so that the red and black wires coming up from the transformer will go through the hole closest to the red post and the blue and black wires go through the hole closest to the black post.

5. Solder the red wire to the outside pin on the red post and the blue wire to the outside pin on the black post. Each black wire goes to the inside pin on each post.

Fourth Type

In a few models, no wires or circuit boards are on top of the video drum.

1. Remove the two mounting screws on top of the drum.

2. Look at the top of the video drum to see how the heads are attached to the video transformer inside.

FIGURE 18.6

Four video head wires soldered to pins.

FIGURE 18.7

Video wires connected to the two small circuit boards on each side of the video transformer.

Video head plugs on top of the drum.

3. In this type of unit, two plugs are on top of the drum, as shown in Figure 18.8. Pull the upper video drum straight up and off.

4. Before remounting the upper drum, read the section in this chapter, "Remounting the upper video drum."

5. Be sure to align the four pins that protrude up from the video transformer (see Figure 18.9) to the plugs in the upper video drum. This type of unit has no colors to align.

Four-headed Machine

These models have a larger round circuit board on top of the upper drum (refer to the section in this chapter, "Second type"). Four-headed units have eight sets of soldered pins, four sets of pins on each side of the circuit board above each video head. The video transformer pins are the two sets furthest away from each video head, as shown in Figure 18.10 (I drew arrows pointing at the four sets of pins). Unsolder these four sets of pins. After unsoldering the pins, pull the upper drum straight up and off.

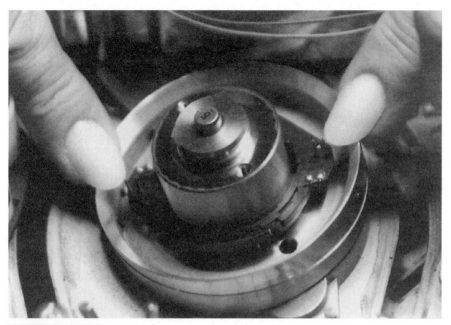

FIGURE 18.9

The pins on top of the video transformer.

FIGURE 18.10

The four sets of soldered pins attached to the video transformer inside the drum.

Before remounting the upper drum, read the section in this chapter, "Remounting the upper video drum." Be sure to align the eight pins protruding up from the video transformer to the pinholes in the circuit board on top of the upper video drum.

Remounting the Upper Video Drum

Wash your hands. Hold the upper drum over the lower drum and check the following:

1. If the video transformer uses soldered pins, check to see if the pins from the transformer are aligned with the pinholes in the circuit board on the upper drum.

2. Check to see if the white part of the circuit board on the upper drum is directly over the white part of the circuit board on the video transformer.

3. If the video transformer has wires attached to it, check to see if the red wire or red post on top of the upper drum is on the same side as the red wire coming up from the video transformer.

4. Align the two screw holes on top of the upper video drum to the two threaded screw holes on top of the transformer, as shown in Figure 18.11.

5. Next, carefully slide the large center hole in the middle of the upper drum over the shaft (refer to Figure 5.2) and slide it down just enough to hold the drum in place.

6. Recheck the screw hole alignment and be sure that the holes in the upper drum are exactly aligned with the threaded holes and that the pins are aligned with the pinholes. If not, remove the drum and start over.

If the alignment is good, push the upper drum down until you can see the threaded screw hole against the holes in the upper drum. Then, replace the mounting screws. Solder the pins or wires back onto the small circuit board on top of the video drum. (Refer to the section in Chapter 17, "Soldering pins to a circuit board.") To solder the wires to the circuit board, read the section in Chapter 31, "Repairing cracks on circuit boards." Then, refer to the section in Chapter 5, "Cleaning the heads."

FIGURE 18.11

The threaded screw holes on top of the video transformer.

Removing Drum Motors

Before proceding in this section, read the section in this chapter on, "The video drum". In newer models, you need to remove the drum motor before removing the video drum.

1. Remove the transport (refer to the section in Chapter 9, "Transports").
2. In models with drum motors mounted to the bottom of the video drum, start by removing the grounding spring on top of the flywheel (refer to Figure 22.3).
3. Remove the two screws on top of the flywheel on the drum motor (refer to Figure 19.12) and remove the flywheel. (The flywheel is hard to lift up because the magnets inside are holding it down.)
4. Remove the three mounting screws to the drum motor and unplug the ribbon, as shown in Figure 18.12.
5. Lift the motor off the lower video drum.
6. On the shaft, under the motor, is a round brass mount where the flywheel was mounted. Measure the distance from the base of the lower video drum to the top of the brass mount, as shown in Figure 18.13. Be sure to write this measurement down.

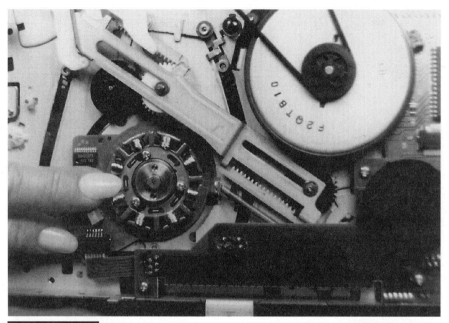

FIGURE 18.12

Three mounting screws and the ribbon plugged into the drum motor.

FIGURE 18.13

Measuring the distance from the base of the lower drum to the top of the brass mount.

7. Loosen the nuts on the side of the brass mount with an Allen wrench.

8. Slide the brass mount up and off the shaft.

When remounting, reverse the process. Be sure to measure and set the brass mount on the new video drum shaft at the same exact distance as you measured before. Now, proceed to the section in this chapter, "The video drum."

In other models, the drum motor is mounted to the top of the upper video drum.

1. Remove the two screws on top of the drum motor and unplug the ribbon, as shown in Figure 18.14.

2. Lift the motor off.

3. Under the motor is a round magnet. Remove the two screws holding the magnet to the upper drum (Figure 18.15) and remove the magnet.

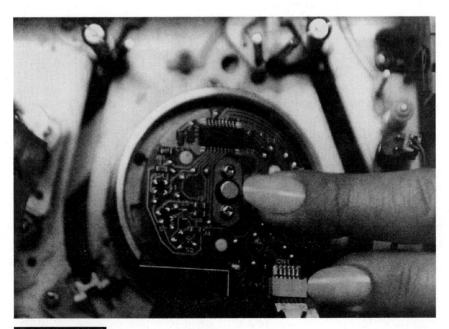

FIGURE 18.14

The mounting screws and ribbon plug on top of the drum motor.

FIGURE 18.15

Mounting screws to the drum motor magnet.

4. On the shaft, under the motor, is a round brass mount where the drum motor was mounted. Measure the distance from the base of the upper video drum to the top of the brass mount (refer to Figure 18.13).

5. Then follow the same procedure on the brass mount as in the above section.

In models that have a rotary cap on top of the upper video drum, refer to the section in Chapter 22, "Third location" under the subhead "Locating the high-pitched sound."

The Video Drum

In newer models, the upper and lower video drums are pressed together at the factory and don't come apart. You can tell you have this type because there will be no Phillips head screws on top of the video drum (refer to Figure 8.17). You will need to replace the entire video drum to replace the video heads.

1. Remove the drum motor. (Refer to the section in this chapter, "Removing drum motors.")

2. Unplug the ribbon protruding out from the lower drum (refer to the section in Chapter 6, "Finding the video leads in newer units").

–In some models, you can reach the three mounting screws at the base of the video drum in the undercarriage, as shown in Figure 18.16. (I drew arrows pointing at the mounting screws.)

–Place the transport on its side. Reach around and hold the video drum.

–Remove the mounting screws and lift the video drum off the transport.

–In other models, you can't reach the mounting screws at the base of the video drum. Instead, remove the bracket holding the video drum to the transport.

–On top of the transport are three mounting screws, as shown in Figure 18.17. (I drew arrows pointing at the mounting screw holes.) Unplug the ribbon protruding out from the lower drum.

–Lift the video drum, the bracket, and the V-mounts straight up and off.

FIGURE 18.16

Three mounting screws at the base of the video drum in the undercarriage.

FIGURE 18.17

The three mounting screw holes on the video drum
mounting bracket.

–Flip the drum over and remove the three mounting screws under
the bracket (refer to Figure 18.16) and remove the bracket from the
drum.

To remount the drum, reverse the process. To order a new video
drum, refer to the section in Chapter 1, "Parts."

Review

1. Wash your hands before removing or replacing the video drum.

2. If the unit has two mounting screws on top of the upper drum, the
   upper video drum comes off.

3. Some units have adjustment nuts on top of the video drum. Do
   not touch these adjustments. They are factory set.

4. When replacing the upper video drum never take hold of the
   drum where the video heads protrude out on the sides of the
   drum.

5. Look at the top of the video drum to see how the heads are attached to the transformer inside.

6. Two-headed machines have four sets of pins and four-headed machines have eight sets of pins soldered to the circuit board on top of the video drum.

7. Only unsolder the transformer pins that are the pins furthest away from each video head.

8. In newer models, the upper and lower video drums are pressed together. You will need to replace the entire video drum.

9. In newer models, you need to remove the drum motor from the video drum.

10. After replacing the video drum, clean the video drum.

Checking and Replacing DC Motors

A VCR contains from two to five separate DC (direct current) motors. This chapter covers the functions and how to identify and locate each motor mounted to the transport. The transport is the foundation where all mechanical parts are connected; it's the chassis of the complete carriage.

Cassette Housing Loading Motors

Cassette housing loading motors drive the cassette carriage and pulls the video cassette into the unit. When you push Eject, this motor drives the cassette carriage to push the video cassette back out. This motor is located in one of four locations:

- In older models, the cassette housing motor is located on the right side of the cassette carriage. Its shaft is directly connected to the worm gear, driving the carriage, as shown in Figure 19.1.

- The second location is on the front of the cassette carriage, driving a worm gear.

FIGURE 19.1

A worm-gear-driven cassette housing motor.

QUICK»»TIP To retrieve either one of these three motors, you must remove the cassette carriage. Refer to Chapter 8, "Removing and servicing cassette carriages."

TRADE SECRET Newer models have no housing loading motor mounted to the cassette carriage to receive or eject the video tape. Instead, this is done by the loading motor on the transport. The cassette carriage has a gear mounted at the bottom of the carriage and is connected to a gear on the transport.

- The third location is at the rear of the cassette carriage. A belt and pulley are attached to its shaft. Figure 19.9 shows the motor mounted horizontally.

- In other models, the location is horizontally set to the transport. This motor has two belts instead of one, as shown in Figure 19.2. This is called a *loading motor* and is also used as a cassette housing motor.

Loading Motors

The loading motor puts the VCR into the Play, Fast Forward, Rewind, Pause, and Stop modes. It also is the vehicle that the

FIGURE 19.2

A belt-driven loading motor.

video tape uses to load or unload the video tape onto the video tape path; in newer models, it receives or ejects the video tape from the cassette carriage. Loading motors have seven different locations.

First Location

In older models, the location is set in a vertical position, (see Figure 19.9). The loading motor is mounted on top of the transport and to the right side of the video drum. Its shaft is pointing down. A shaft is the part of the motor that rotates and has a pulley or worm gear attached to it. The belt and pulley are in the undercarriage.

Second Location

In other models, the location of the loading motor is horizontally set on top of the transport. A belt and a pulley are attached to its shaft. Its shaft is pointing to the rear of the unit and is mounted either on the right side of the transport, as shown in Figure 19.3, or mounted the same way on the left side of the transport.

FIGURE 19.3

Second location of a loading motor.

Third Location

In some models, the location of the loading motor is horizontally set as well, on top of the transport and has two belts and a double pulley attached to its shaft. Its shaft is pointing to the front of the unit and is mounted to the transport on the right side of the video drum (see Figure 19.2).

Fourth Location

In a few models, the location is horizontally set and is mounted in the undercarriage. Part of the loading motor protrudes through a hole in the transport, as shown in Figure 19.4. It has a belt and pulley attached to its shaft. The belt drives a worm gear in the undercarriage. Its shaft points toward the back of the unit. In other models, the shaft points toward the left side of the unit and has a worm gear attached to it.

Fifth Location

In newer models, the location is horizontally set on top of the housing bracket, which is on top of the transport and has a worm gear attached

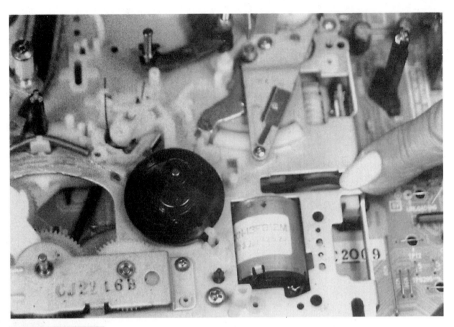

Fourth location of a loading motor.

to its shaft. The shaft might point toward the front, rear, or to the left side of the unit, and the loading motor is mounted on top of the housing on the right side of the video drum, as shown in Figure 19.5.

Sixth Location

In other models, the location is horizontally set as well and is mounted on the right or left side in the undercarriage. This loading motor can have a worm gear attached to the shaft or a belt and pulley. Its shaft can be pointed toward the rear or to either side of the unit (see Figures 19.21 and 19.22). Refer to Chapter 9, "Getting into the undercarriage."

Seventh Location

In newer models, the location is horizontally set underneath the transport and has a worm gear attached to its shaft. The shaft usually points toward either side of the unit and the loading motor is mounted in the middle or off to the right side in the front of the transport in the un-

FIGURE 19.5

Fifth location of a loading motor.

dercarriage, as shown in Figure 19.6. Refer to the section in Chapter 9, "Transports."

Reel Motors

Reel motors are used to drive the take-up spindle, which pulls the video tape back into the video cassette when in the Play mode. This motor also drives the supply spindle for Fast Forward and take-up spindle for rewind. There are two locations for reel motors.

A few models have two reel motors that are located directly under the take-up and supply spindles in the undercarriage, as shown in Figure 19.7. The spindles on top of the transport are attached directly to the shafts of the motors. The motor on the right is called the *supply reel motor* and the motor on the left is called the *take-up reel motor*.

Other models have only one reel motor, rather than two. This motor is located under and between the two spindles in the undercar-

FIGURE 19.6

Seventh location of a loading motor.

FIGURE 19.7

The take-up reel motor (left) and supply reel motor (right).

FIGURE 19.8

A reel motor.

riage, as shown in Figure 19.8. This motor drives all three functions (fast forward, rewind, and take-up) with the aid of an idler wheel (refer to Figure 24.3). The idler wheel moves from spindle to spindle, depending on which direction the motor is turning.

Capstan Motors

The capstan motor drives the capstan shaft. The capstan shaft pulls the video tape through the video-tape path. If the unit has no reel motor, then the capstan motor drives the take-up spindle, which pulls the video tape back into the video cassette when in the Play mode. This motor also drives the supply spindle and take-up spindle for Fast-Forward and Rewind modes. For more information, refer to Chapter 25, "Capstan shaft problems."

Belt-driven Capstan Motors

Older models have one location to find the belt-driven capstan motor. The location is to the right of the video drum (refer to Figure 4.7). It's

FIGURE 19.9

Cassette housing motor (bottom), loading motor (left), and capstan motor (right).

the largest vertically mounted motor with a flywheel on top of it. Its belts and pulley are located in the undercarriage. One to three drive belts are on its pulley.

Study Figure 19.9. It shows a combination of three different motors mounted to the right of the video drum. The motor mounted horizontally is the cassette housing motor. The motor to the right, mounted vertically with a flywheel on top of it, is the capstan motor. The motor to the left, mounted vertically, is the loading motor.

Direct-Drive Capstan Motors

Newer models use a direct-drive capstan motor directly under the capstan shaft in the undercarriage. The actual base, or flywheel, of the capstan shaft is part of the motor. This flywheel is the largest wheel in the undercarriage.

To identify a direct-drive capstan motor, locate the circuit board that protrudes from underneath the large flywheel, as shown in Figure 19.10. In some models, the circuit board completely surrounds the flywheel, as shown in Figure 19.11. For more information, refer to the section in Chapter 25, "Removing the capstan shaft in direct drive VCRs."

FIGURE 19.10

A circuit board protruding out under the flywheel.

FIGURE 19.11

A circuit board completely surrounding the flywheel.

In some models, the direct-drive motor has a reel drive belt attached to a small pulley in the center of the flywheel (refer to Figure 10.2). This belt drives the take-up spindle, fast forward, and rewind positions. Don't get this system confused with a belt-driven system. If you have a belt-driven system, the belt is around the outside edge of the flywheel (refer to Figure 25.4).

Drum Motors

The drum motor drives the upper video drum. The upper drum has the video heads attached to it. The video heads read the image off the video tape. The drum motor has two possible locations. In older models, the round cylinder attached to the lower video drum is the motor (refer to the base, left side of Figure 5.1). In newer models, the motor is located directly under the lower video drum in the undercarriage and is a direct-drive motor, as shown in Figure 19.12. In other models, the motor is located directly on top of the upper video drum and some-

FIGURE 19.12

Direct drive drum motor.

QUICK»»TIP If the drum motor is making a funny-sounding noise while it's running or if it sounds like something is dragging inside the drum, check for a piece of video tape caught between the upper and lower video drums. To remove the upper video drum, refer to Chapter 18, "Replacing the video heads."

times called the *rotary cap* (refer to Figure 22.4). This rotary cap is part of the direct-drive motor. For more information, refer to the section in Chapter 18, "Removing drum motors."

Checking Motors for Dead Spots

If a motor doesn't start when you push the corresponding mode button (see chart below), a motor coil might have opened. An open coil is a dead spot. A coil consists of wire wrapped around metal plates inside the motor. The metal plates turn into an electrical magnet when electricity flows through the coil.

To check for a dead spot, locate the spindle, pulley, or worm gear that is directly connected to the shaft of the motor. Push the corresponding mode button, then spin the pulley or worm gear in either direction immediately afterwards. If the motor starts operating on its own, it has a dead spot. The motor will continue working until it stops on a dead coil. If the motor has a dead spot, it must be replaced.

Which mode button activates the correct motor?

Motor	Button
Cassette housing	Eject or insert a video cassette
Loading	Play or Stop
Capstan	Play, Fast Forward, or Rewind
Reel	Rewind or Fast Forward
Drum	Play

Checking for Dead Motors

If you can't spin the pulley when pushing the correct mode button, remove the drive belt from the pulley attached to the motor shaft. Then, push the corresponding mode button to see if the motor runs without a belt. If the motor doesn't run after removing the belt off the pulley, then check for a dead spot, bad bearings, or an open fuse.

If a motor has a worm gear attached to its shaft and you can't turn the gear while pushing the corresponding mode button, remove the motor, leaving the leads to the motor attached. Then, push the correct mode button to see if the motor runs. If it doesn't run, check for a dead spot, bad bearings, or an open fuse.

If the motor does run, check all the connecting moving parts that the belt or worm gear drives. A bent bracket, lever, or a foreign object might be jamming the unit or the gearing could be out of alignment. Refer to Chapter 20, "Mechanical" under the subhead "Diagnosing VCR problems," Chapter 15, "Gear alignment," Chapter 25, "Detecting a capstan shaft problem" under the subhead "Capstan shaft problems" and Chapter 26, "Jammed cassette carriages."

Checking for Bad Bearings

If the shaft of the motor doesn't turn or turns very slowly, follow these steps:

1. Unplug the unit and remove the motor drive belt.
2. If a worm gear is attached to the motor, remove the motor.
3. Spin the pulley or worm gear on the shaft of the motor.
4. Check the pulley or worm gear on the shaft of the motor in both directions to see if it turns freely. If one bearing is bad, the shaft will be hard to turn or will be moving unevenly. If all the bearings are bad, it will be hard to turn or won't turn at all. If this is the case, the bearings are bad and the motor must be replaced.

There is another way to check for bad bearings.

1. In motors that have a belt attached to it, remove the belt.
2. In motors that have a worm gear attached to it, leave the motor in place.
3. Plug the unit back in.
4. Push the corresponding mode button. If you hear a loud squeal when the motor runs, the bearings are bad and the motor must be replaced.
5. For replacement of the motor, refer to the section in Chapter 1, "Parts."

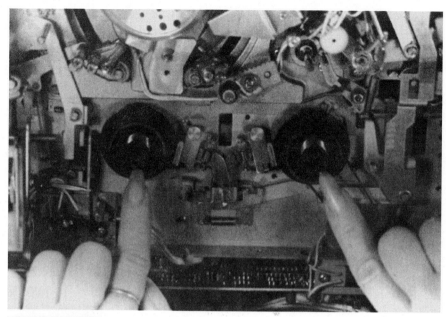

FIGURE 19.13

Supply spindle (left), and take-up spindle (right).

Checking Reel Motors

To check a reel motor without getting into the undercarriage, look between the two spindles. This type of system has no pulleys, wheels, or
gears between the two spindles. All you'll find is a brake shoe assembly. The two spindles are directly connected by a shaft to the reel motors, as shown in Figure 19.13. Reel motors control the fast forward,
rewind, and take-up functions.

1. Insert a blank cartridge (refer to Figure 1.1).
2. Push Rewind.
3. Check the supply spindle on the left to see if it's turning. If not, give
   the spindle a quick spin counterclockwise. If the spindle operates
   on its own, it has a dead spot in the motor and the motor has to be
   replaced.
4. If the spindle doesn't operate on its own, push Stop and look to the
   right side of the spindle for a brake shoe.

Pulling the brake shoe away from the spindle.

    A. Pull the brake shoe away from the spindle, as shown in Figure 19.14.

    B. While holding back the brake shoe, turn the supply spindle to see if it spins freely. If the spindle is hard to turn, the bearings are bad and the motor has to be replaced.

    C. If neither of these instances are the case, the problem is the circuitry. If you have a circuit problem take your VCR to your local service center for repairs.

5. Locate the take-up spindle on the right.

6. Push Fast Forward.

7. If the take-up spindle doesn't turn, give it a quick spin clockwise.

8. If it starts turning, the motor has a dead spot and has to be replaced.

    Check for bad bearings by following the same procedure explained in the section above.

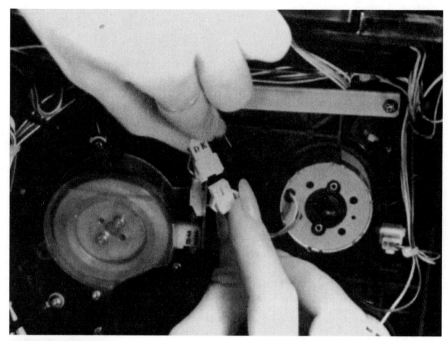

FIGURE 19.15

Unplugging the motor.

Removing a Reel Motor

1. Unplug the VCR and remove the cassette carriage. Refer to Chapter 8, "Removing and servicing cassette carriages," and refer to the section in Chapter 28, "Removing spindles attached to a motor."

2. Place the unit on its side.

3. Remove the bottom cover plate and open up the circuit board to the undercarriage. Refer to Chapter 9, "Getting into the undercarriage."

4. You now have a clear view of the take-up and supply motors. Refer to Figure 19.7 for the location of these motors. The base of each motor has two separate wires attached. Follow the wire coming from the reel motor to a plug, as shown in Figure 19.15. Unplug the plug.

    Some models have no plug at the end of the wires connected to the reel motor. Instead, the wires are soldered onto terminals on the back of the motor. You'll find a small wire soldered to each terminal. Each wire will be a different color. Before detaching the wires, read the section in this Chapter, "Soldering wires and terminals." Then,

FIGURE 19.16

Removing a reel motor.

cut the wires off close to each terminal with a pair of wire cutters. These wires have to be re-soldered when you replace the motor.

5. Now, you're ready to remove the motor. Leave the unit on its side. Remove the two or three Phillips head screws located on top of the transport under the spindle that you removed previously. At the same time, you need to take hold of the motor in the undercarriage to keep the motor from falling out when you remove the last screw, as shown in Figure 19.16.

6. Pull the motor straight out from the undercarriage side.

7. Take this motor with you to your local electronic supply store or refer to the section in Chapter 1, "Parts."

To install the new motor, simply reverse the process.

In models with one reel motor (see Figure 19.8), follow the same procedure for removing a reel motor. But in this case:

1. Instead of removing a spindle, remove the two Phillips head mounting screws, one on each side of the plastic bracket (refer to Figure 24.17).

2. Remove the spring attached between the idler wheel and the plastic cover.

3. Remove the plastic cover and the motor.

4. After removing the motor, you will see a small wheel attached to the shaft.

5. Loosen the lock nut at the base of the wheel with a 1.5-mm Allen wrench and slide the wheel off the shaft.

To install the new motor, simply reverse the process.

In a few models, the reel motor has a gear directly connected to the reel motor shaft. This gear drives the take-up spindle. Two Phillips head screws hold the motor in place, one on each side of the gear, as shown in Figure 19.17.

1. With a flathead screwdriver, pry the gear off the motor shaft.

2. Put the gear aside; you'll need it for reassembling.

FIGURE 19.17

Aligning the teeth of the gear to the take-up spindle.

Follow the same procedure as previously explained to replace the reel motor.

After replacing the new motor, turn the unit around and replace the gear that you previously removed. Slide the gear down the shaft until the teeth on the gear are perfectly aligned to the teeth on the take-up spindle (see Figure 19.17).

Removing Loading Motors and Belts

Unplug the unit. Read the following ways to remove motors and belts, then choose the one that resembles your unit. After removing the loading motor, you need to remove a worm gear or a belt and pulley attached to its shaft. All worm gears slip right off and all pulleys pry off the shaft. You might need to remove the cassette carriage from some units (refer to Chapter 8, "Removing and servicing cassette carriages").

First Type

In this model:

1. Remove the cassette carriage. You can remove both belts without removing the loading motor.
2. The motor is mounted with clips. To remove, spread the clips apart and pry the motor out with a flathead screwdriver.
3. This motor is mounted horizontally (see Figure 19.2). The wires are soldered onto terminals on the back of the motor.
4. A small wire is soldered to each terminal. Each wire is a different color.
5. Before detaching the wires, read the section in this chapter, "Soldering wires and terminals." Then, cut the wires off close to each terminal with a pair of wire cutters.

To install the new motor, place the motor over the clip and push the motor down into the clip until the clip snaps around the motor.

Second Type

In this model:

1. Remove the cassette carriage. You cannot remove the belt without removing the loading motor first.

2. Remove the three mounting screws on top of the housing below the motor (refer to Figure 27.26).

3. Remove the belt off the motor pulley.

4. Unplug the motor and lift the housing and motor straight up and out.

5. Remove the motor from the housing by removing the two screws attached to the front of the motor, one on each side of the shaft.

To install the new motor, simply reverse the process.
To replace the belt:

1. Remove the O-ring above the pinch-roller bracket (see Figure 27.27) and pull the bracket straight up and off. In this case, the tension spring is incorporated within the bracket itself and comes off with the bracket.

2. The tracking arm will come off the cam gear. Be sure that the tracking arm is back in the right groove when replacing it.

3. Remove the cam gear by lifting it straight up and off.

4. Remove the worm gear, pulley, and belt. Be sure to note how it comes out.

5. Place the new belt around the drive pulley.

6. Place the worm gear back onto the transport.

To remount the cam gear, refer to the section in Chapter 15, "Main cam gear alignment." To remount the pinch-roller bracket, simply reverse the process or refer to the section in Chapter 27, "Removal of pinch-roller brackets."

Third Type

In this model:

1. Remove the cassette carriage. You can remove all three belts without removing the loading motor.

2. Remove the two mounting screws at the base of the housing bracket, as shown in Figure 19.18.

3. Remove the belt at the rear of the housing.

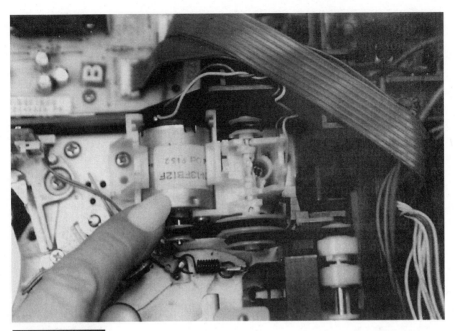

The loading motor and the two housing mounting screws.

4. Lift the housing and motor straight up and out.

5. Remove the belt from the motor pulley.

6. Remove the two screws attached to the front of the motor, one on each side of the shaft.

7. Slide the motor out of the housing. Before detaching the wires, read the section in this chapter, "Soldering wires and terminals."

8. Cut the wires off close to each terminal with a pair of wire cutters.

To install the new motor, simply reverse the process.

Fourth Type

This model uses a worm gear.

1. To remove the loading motor, pick up on the back of the plastic clip at the rear of the motor, as shown in Figure 19.19.

2. While lifting up on the clip, pull the clip straight back until the clip comes off.

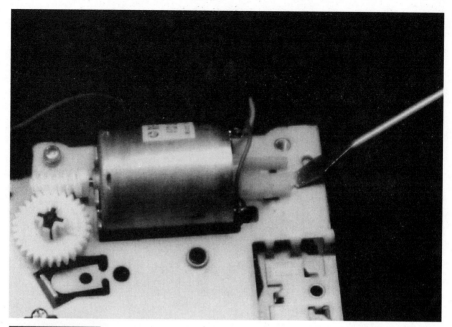

FIGURE 19.19

Picking up on the clip to release the loading motor.

3. Put the clip aside, you'll need it for reassembling.

4. Pull the motor straight back and then lift it straight up.

5. Before detaching the wires, read the section in this chapter, "Soldering wires and terminals."

6. Cut the wires off close to each terminal with a pair of wire cutters.

To install the new motor, note the two small pins on top of the housing bracket, as shown in Figure 19.20. Place the two threaded holes in the front of the motor over each pin, one on each side of the shaft. Next, place the plastic clip into the clip hole behind the motor and push forward until it snaps into place.

Fifth Type

This model also uses a worm gear.

1. Remove the loading-motor housing bracket, which covers the motor.

2. Remove the three mounting screws in the housing (refer to Figure 15.6).

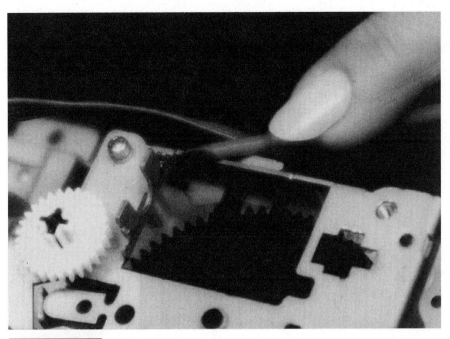

FIGURE 19.20

Pins on top of the housing bracket.

3. At the back of the motor, a small circuit board is plugged into the mother board. Lift the housing and motor straight up and the plug behind the motor will automatically unplug.

4. Flip the housing over. Inside the housing, you will see the loading motor, a worm gear, and worm wheel (refer to Figure 15.7).

5. A clip is on each side of the motor. Spread the clips apart and pull the motor and small circuit board straight out of the housing. If the worm wheel falls out, don't worry, just put it back in place. You do not need to align this gear. Refer to the section in this chapter, "Soldering wires and terminals."

To install the new motor, place the motor over the clip and push the motor down until the clip snaps around the motor. Remount the housing bracket by placing the housing over the pinch roller cam gear and aligning the screw holes. Then, push the bracket straight down. The plug at the back of the housing will plug right into the mother board behind the transport. Remount the mounting screws.

FIGURE 19.21

Removing the belt and mounting screw to remove the loading motor.

Sixth Type

In this model:

1. Place the unit on its side.

2. Remove the bottom cover plate. Refer to Chapter 9, "Getting into the undercarriage."

3. You can remove the belt to the left without removing the loading motor. To replace the belt on the right, the motor needs to be removed first, as shown in Figure 19.21. There is a housing bracket under the motor with one screw.

4. Pull the belt off the motor pulley and remove the screw.

5. Lift up on the side of the housing that had the screw in it and slide the motor out.

6. At the rear of the motor, a little circuit board has a plug on it. Unplug the motor.

7. Remove the two screws in front of the motor, one on each side of the shaft, then slide the motor out of the housing.

8. To remove the small circuit board from the motor, refer to the section in this chapter, "Soldering wires and terminals."

To install the new motor, mount the motor to the housing, position the housing so that you can slide the sleeve of the housing, on the opposite side of the screw hole, into a slot in the transport. Push the motor back into place and remount the screw.

Seventh Type

In this model:

1. Place the unit on its side.

2. Remove the bottom cover plate (refer to Chapter 9, "Getting into the undercarriage").

3. In this VCR, it appears that you can't replace the belt without removing the loading motor, but you can. There are three mounting screws, as shown in Figure 19.22. Remove the belt and the three screws.

4. Lift the housing and motor straight up and out.

5. To remove the shielded wire from the back of the motor, refer to the section in this chapter, "Soldering wires and terminals."

6. Remove the two screws on the front of the motor, one on each side of the shaft, and slide the motor out of the housing.

To install the new motor, simply reverse the process.

Eighth Type

In this mode:

1. Remove the transport. Refer to the section in Chapter 9, "Transports." This model uses a worm gear (see Figure 19.6).

2. At the back of the loading motor, where the wires are attached, is a latch. Place a flathead screwdriver under the rear of the motor. Pull back on the latch and simultaneously pry the motor up and out. The worm gear will come right off the motor.

3. Before detaching the wires, read the section in this chapter, "Soldering wires and terminals."

4. Cut the wires off close to each terminal with a pair of wire cutters.

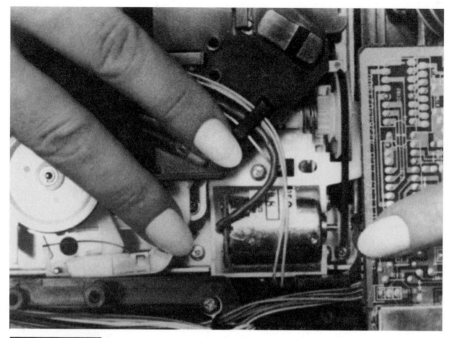

FIGURE 19.22

Removing the three screws to remove the loading motor.

To install the new motor, simply reverse the process.

Ninth Type

In this model:

1. Remove the cassette carriage and the transport.

2. In this type of unit, you can remove the belt without removing the loading motor. The motor is mounted directly to the transport in the undercarriage (refer to Figure 19.4).

3. Remove the motor by removing the two screws attached to the front of the motor in the undercarriage, one on each side of the shaft.

4. Lift the motor straight up and out.

5. Before detaching the wires, read the section in this chapter, "Soldering wires and terminals."

6. Cut the wires off close to each terminal with a pair of wire cutters.

To install the new motor, simply reverse the process.

Replacing All Other Motors

All motors have either a worm gear, a drive wheel, a spindle, or a belt and pulley attached to its shaft. All worm gears slip right off. All drive wheels are attached with a locking nut. Most pulleys pry off of the shaft. If the piece doesn't come off easily, look for lock nuts located on the base of the gear, wheel, or pulley. With a 1.5-mm Allen wrench, loosen these lock nuts to remove the part.

Locate the wires attached to the back of the motor. Follow the wires to a plug and unplug it. If there's no plug, cut the wires off and resolder them. Refer to the section in this chapter, "Soldering wires and terminals."

All motors are mounted by two or three Phillips head mounting screws. If the motor is mounted to the bottom of the transport in the undercarriage, the mounting screws come down through the top of the transport to the motor. If the motor is on top of the transport, the mounting screws are in the undercarriage. All other motors are horizontally mounted to a housing bracket. The mounting screws are on one side of the housing and the motor is on the other. In some models, you'll have to remove the housing to remove the motor. Locate the screws on the housing bracket and remove them. Now, remove the housing and the motor with it. Then, remove the motor from the housing.

All direct-drive motors, as in Figure 19.10 and Figure 19.11, have their own circuitry as part of the motor. If the motor doesn't run, it is caused by a circuit problem either in the motor itself or a circuit problem in the unit going to the motor. You need to take the VCR into your local service center for repair.

Soldering Wires and Terminals

A small wire is soldered to each terminal. Each wire will be a different color. A plus and a minus sign is beside each terminal, as shown in Figure 19.23. Draw a diagram of what colored wire goes to which marked terminal. This diagram will keep you from confusing the wires when reassembling the VCR.

After installing the new motor, you'll need to solder the wires back to the terminals on the back of the motor:

1. With wire cutters, strip off about ¼ inch of the insulation covering the wires.

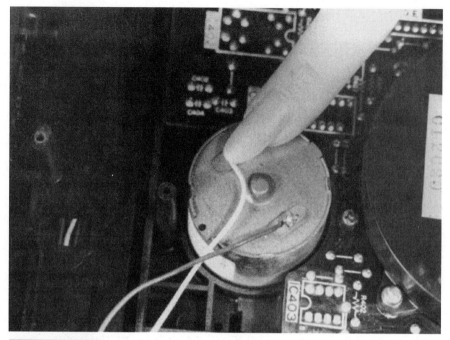

FIGURE 19.23

The plus and minus sign beside each terminal.

2. Put the correct colored wire through the correct terminal hole and wrap the wire around the terminal.

3. Coat the tip of a soldering iron with solder.

4. Flip off any excess solder.

5. Touch the tip of the soldering iron and the solder directly to the terminal to form a bead of solder on the terminal, as shown in Figure 19.24.

6. Remove the solder, leaving the soldering iron in place for a few seconds to bond the wire to the terminal.

Repeat this procedure for the other terminal.

For motors that have a small circuit board attached to the back of the motor, you need to remove the circuit board. Refer to the sections in Chapter 17, "Using solder wick to remove old solder" and "Soldering pins to a circuit board." A plus and minus sign is printed

FIGURE 19.24

Soldering a wire onto a terminal.

on the circuit board where the pins from the motor protrude through the circuit board. Also, a plus and minus sign is beside each pin on the back of the motor (see Figure 19.23). Align the plus and minus signs on the motor to the signs on the circuit board.

Review

1. If the motor doesn't run, push the correct mode button and try spinning it to get it started to check for dead spots.

2. Be sure that you push the proper mode button to activate the correct motor.

3. Check the motor for bad bearings.

4. If the motor shaft doesn't move at all, disconnect the shaft from any belt or gear. If it starts running:

   A. Check for a foreign object jamming the gearing.

   B. Check for a bent bracket or lever jamming the gearing.

   C. Check if the gearing is out of alignment.

5. All motors are mounted by two or three Phillips head screws.

6. All motors are mounted on the opposite side of the transport or opposite side of the bracket where the motor is located.

7. When replacing the DC motor, be sure to replace the correct wire to the correct terminal on the motor.

8. When remounting the small circuit board to the motor, be sure that the polarity is correct.

Diagnosing VCR Problems

This chapter covers the procedures for diagnosing problems in a VCR. The problems have been grouped into three major categories, based upon the type of symptoms that they exhibit: video, audio, or mechanical. You'll be referred to the proper chapter for instructions on how to correct each particular problem.

Video

Many of the problems that your VCR develop are noted because of the effects they will have on the picture that you see on your television. Choose the section that describes the problem that the VCR is experiencing and follow the instructions describing how to repair the problem.

Distorted Picture

While you are viewing videotape, the picture becomes partially or completely distorted or snowy. The TV screen might even go completely black or blue; however, the audio portion is okay.

There are two ways to check out this problem.

1. If you have a stereo hi-fi unit and the audio is distorted, place the stereo mode switch to the Normal position.

2. Listen to hear if the audio clears up. If it does, you could have a dirty video head.

3. Switch the VCR/TV switch to the TV position and tune in a broadcast TV station. If the program comes in clear on the TV monitor, you have eliminated the circuitry and it probably is a dirty video head. Refer back to Chapter 5, "Cleaning the video heads" and to Chapter 6, "Pinpointing and correcting video head problems." If ¼ to ⅔ of the screen is distorted or snowy, but the remainder of the screen is clear, proceed to Chapter 12, "Video tape path alignment."

Lines Following an Object

You see black or white trailing lines shooting off the right side of an object in the picture. They are on the entire screen. The more pronounced the object, the more visible the lines.

The cause of this problem is a bad video head. If trailing lines still appear after cleaning and saving the video heads, then the video head is bad and has to be replaced. Refer back to Chapter 18, "Replacing video heads."

You can make other adjustments to help clear the picture up. Refer to the sections in Chapter 11 on "Back tension guide adjustment" and "Brake shoe band adjustment." Then refer to the section in Chapter 14, "FM Alignment."

Stationary Lines on Top or Bottom of the Picture

When you play any video tape, you find a group of horizontal lines across the top or bottom of the screen. These groups of lines are stationary (they do not float through the picture). You also notice that the tracking adjustment moves these lines up or down the screen, but they still remain on the screen. Many groups of horizontal lines might run across the screen, covering nearly the entire picture.

QUICK»»TIP If the lines can be adjusted out of the picture with the tracking control, the video heads are on the borderline of going bad.

If either of these conditions exist, the videotape path is out of alignment. Proceed to Chapter 12, "Video tape path alignment."

While you are trying to view a movie, a horizontal belt of lines appears on your TV screen. You find the tracking adjustment has very little effect. Remove the top cover and you might find a foreign object like a straw or piece of paper caught between one of the roller guides and the V-mount. Refer to the section in Chapter 12, "Identifying roller guides and V-mounts." Just a small piece of paper caught behind one of the roller guides can throw the guide out of alignment. Simply remove the paper.

Lines Appear in the Picture in Some Tapes

You have to adjust the tracking control every time you play videotape to remove the lines from the picture.

Re-center the tracking control. Proceed to Chapter 14, "FM alignment."

Jitter in the Picture

You're viewing a videotape and there's a jitter (the picture jumps up and down) in the picture. You adjust the tracking control on the VCR and check the vertical hold control on the TV monitor, but these adjustments have little or no effect.

Proceed to the section in Chapter 12, "Checking the adjustment" and "Fine tuning the roller guides." You also should check the torque on the videotape. If the picture is okay in the two-hour mode, but has a jitter in the six-hour mode, proceed to the section in Chapter 11, "Back-tension guide adjustment." Adjust the back-tension guide using a six-hour videotape.

Lines Floating Through the Picture

You view videotape and you see a belt of lines moving continuously from the bottom to the top of the screen. The lines also can be floating from top and bottom and be meeting in the middle.

QUICK»TIP

Another way to diagnose this problem is to place your finger on top of the back-tension guide pole, then push the guide pole to the left while the unit is playing. If this adjustment clears the picture, proceed to Chapter 11, "Setting the torque on a video tape." Either the back tension isn't tight enough or the audio head is out of alignment. If neither of these adjustments clears the picture, the problem is a bad A/C head or a circuit failure. You'll have to take your VCR in for service.

FIGURE 20.1

Pushing the bottom portion of the video tape against the A/C head with a glass brush.

The FM portion of the A/C head being out of alignment usually causes this condition. To check for this problem, use a fiberglass brush to push the bottom portion of videotape up against the A/C head while the unit is playing, as shown in Figure 20.1. Using the brush, move the videotape up slightly. If moving the tape up doesn't clear the picture, then push the videotape down slightly. If either position clears the picture, then do an audio head alignment. This alignment will automatically align the FM portion of the A/C head and remove all the lines. Proceed to Chapter 13, "Audio head alignment," and read "Adjusting the FM portion."

Picture Jumping Up and Down

A small group of lines runs across the bottom portion of the screen, or the whole picture jumps up and down. The vertical hold control on

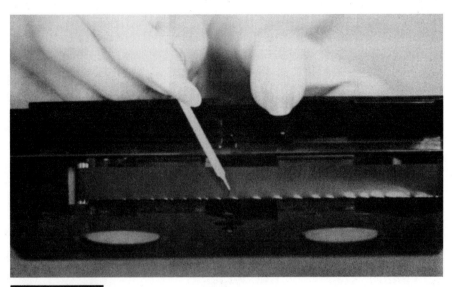

FIGURE 20.2

A crinkle on the bottom portion of the video tape.

the TV monitor and the tracking control on the VCR won't remedy either condition.

Stop the VCR and eject the videocassette. Locate the button on the front left side of the videocassette. Push this button to open the cassette door, exposing the videotape. At the bottom edge you will see a crinkle, as shown in Figure 20.2, which was caused from the tape previously being "eaten." If this problem exists, proceed to the section in Chapter 27, "Second location" under the subhead "Two locations at which a video tape can be 'eaten.'"

Lines and Dashes Across the Screen

Many stationary lines and dashes shooting across the screen. The audio portion is distorted or has a buzz in it.

This problem is caused by the back-tension guide pole being out of adjustment. Proceed to Chapter 11, "Setting the torque on a video tape."

A single, long white or black line shoots across the screen like a star. Another line might also run across the top, middle, or bottom of the screen at the same time. The lines flash sporadically anywhere on the screen, then disappear.

A video head going bad usually causes this condition, called *video dropout*. Refer to Chapter 5, "Cleaning the video heads," and Chapter 6, "Pinpointing and correcting video head problems." If needed, proceed to Chapter 18, "Replacing the video heads."

White dashes shoot across the screen, but the audio portion is okay.

The problem is a grounding leaf spring that removes static electricity from the video drum. If the spring is improperly cleaned or bent, white dashes appear in the picture. Refer to Chapter 22, "High-pitched sound problem." If you can adjust the lines out of the picture by using the tracking adjustment, the FM adjustment is out of alignment. Refer to the sections in Chapter 14, "Locating the FM adjustment" and "Centering the tracking control."

Fluctuating Distortion in the Picture

The picture fluctuates from being stable and clear to being unstable, with a distorted picture. Each cycle takes 5 to 15 seconds. The audio portion sounds like it's speeding up and then slowing down. The audio portion also might be distorted and then clear up simultaneously with the picture.

Proceed to Chapter 11, "Setting the torque on a video tape," or to the section in Chapter 13 on "Adjusting the FM portion." The FM synchronization on the videotape isn't aligned with the A/C head properly, which causes this problem.

Multiple Floating Lines

Periodically, you might see various sizes of lines floating through the picture. Along with this symptom, you'll hear a buzz in the audio.

If this condition occurs while viewing a particular tape, stop the VCR immediately. Remove the videocassette, open the videocassette door, and look at the tape. In this particular case, you'll find that the videotape has been "eaten" across its whole width, as shown in Figure 20.3. Refer to Chapter 27, "Video-tape-'eating' VCRs."

A Bar of White Rectangular and Square Boxes

A configuration of white rectangular and square boxes that resembles a large city appears on the TV. The configuration is a long and narrow

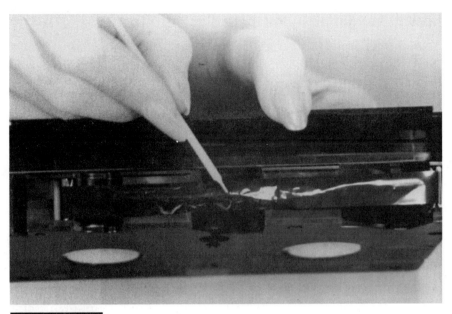

FIGURE 20.3

A video tape eaten completely across.

bar across the screen. It repeatedly starts at the top of the screen, works its way down and disappears. It also might remain stationary at the top of the screen, appearing and disappearing.

Refer to the section in Chapter 11, "Detecting a back-tension guide problem."

A Few Floating Lines and the VCR Stops

The VCR is playing and a few lines suddenly appear on the screen. Then, the unit automatically stops or in some models, goes into rewind. It repeats the process if you push Play again.

The VCR stops at the same point every time because a small pinhole is in the videotape. When the pinhole passes the tape sensors, the sensors signal the unit to stop. So, the problem isn't the VCR. To correct this problem, push Fast Forward for a second or two, push Stop, and then push Play. This will bypass the pinhole. For a permanent solution, remove the videocassette and repair it. Proceed to Chapter 29, "Repairing a video tape."

Speed Fluctuation in Playback Causing Lines

You have recorded a program from a broadcasted TV station. You believe it recorded just fine, but when playing the recording, you see and hear the picture and audio fluctuating.

The VCR doesn't cause this, rather by the antenna system interfering with the TV signal. A distorted cable system can also cause this interference. The interference blocks the internal speed governor signal, which determines the proper tape speed. The interference causes the unit to continuously search for the right speed during playback, causing the unit to speed up and slow down.

I've found that many people use improper antenna systems. The worst antenna system is rabbit ears. This type of antenna picks up interference, even from someone walking too close to it, causing the TV signal to be weak or ghosts to appear in the picture. A weak signal means that only a weak sync pulse reaches the unit; consequently, the unit can't determine at what speed it should run. The following outside antenna systems cause the same interference as rabbit ears:

- An antenna with broken or bent arms.
- An antenna that is not attached properly and flops around.
- An antenna aimed in the wrong direction.
- Using bad lead-in wire.
- Using a plain piece of wire for an antenna.
- A weak cable system.

Audio

If the audio portion of your videotapes begin to sound strange, the problem can be traced to one of many possibilities. Read these sections and follow the instructions to repair your VCR's particular problem.

Distorted Audio

Upon playback, the VCR is producing a good picture, but the audio is distorted or is coming in very weak.

Here are five explanations why this situation can happen:

- A buildup of residue on the face of the audio head from dirty video tapes can cause audio distortion. This buildup prevents the videotape from making a flush contact with the audio head, giving a weak or distorted audio signal. Refer to the sections in Chapter 4, "A/C head and eraser head" under the subhead "Cleaning the various components."

- Improper alignment is another cause of audio distortion. Proceed to Chapter 13, "Audio head alignment."

- The third cause is a bad audio head or a circuit problem. Proceed to the sections in Chapter 17, "Checking for a bad audio head."

- The last two causes occur only in stereo hi-fi models. If the unit is in the stereo hi-fi mode and the audio is distorted or makes a motorboat sound, this distortion might be caused by the tracking control being set improperly. To remedy this complication, readjust the tracking control until the audio becomes clear.

- Another reason that stereo hi-fi units might have distortion is that the video tape path might be out of alignment. Proceed to Chapter 12, "Video tape path alignment." The video heads produces the audio in these types of models. If the tape path is out of alignment, the audio is distorted or has a buzzing sound in the background. If the audio is clear on the stereo hi-fi mode, but is distorted in the Normal position, check for a bad or dirty audio head.

Wavering in the Audio

You hear a wavering in the audio when playing video tape. It's more noticeable in the six-hour mode than the two-hour mode. You also hear this wavering in all commercial video tapes. You'll notice that if you play the recording on another VCR, the wavering sound is still there.

A bad clutch assembly causes the problem. Proceed to the sections in Chapter 23, checking "Clutch assemblies" and "Removing clutch assemblies," and the section in Chapter 24, "Removing idler wheels and clutch assemblies." The type of clutch that usually causes this problem resembles two gears sandwiched together and is located in the undercarriage. To order a new clutch assembly, refer to the section in Chapter 1, "Parts."

Mechanical

The final category of VCR repair is mechanical. These problems don't display symptoms on the television and are not audible. Look through the following sections and find the heading that describes your problem. Follow the instructions in that section.

VCR Shutting Down

You push Play. After 60 seconds, the unit completely shuts down with the tape inside. You've tried pushing the Power button, but with no effect.

Try unplugging the unit and plugging it back in. The microcomputer will reset. When you plug the unit in, it'll fire up right away or you'll have to push the Power button. You'll hear a motor running inside the unit. This sound continues for about 60 seconds, then the unit shuts down. The cause of this problem is a broken loading belt. First, refer to the sections in Chapter 19, "Loading motors" and "Removing loading motors and belts." After locating the loading motor, read the first four sections of Chapter 10.

You insert a videocassette and after 60 seconds the unit shuts down. Unplug the power, then plug it back in to reset the microcomputer. The videocassette ejects within 10 seconds. You insert a videocassette again and it shuts down.

Look inside through the cassette door on front of the VCR. See if the videocassette has loaded all the way. If not, the gear on the cassette carriage is broken. Refer to the section in Chapter 26, "Fifth reason" under the subhead "The reasons why the cassette carriage jams."

You insert a videocassette and push Play. You can hear the VCR loading the video tape onto the tape path; before a picture is present on the TV monitor, it unloads and then stops or shuts down. On the other hand, it might show a picture for a few seconds, it unloads, and then stops or shuts down.

The problem is a bad mode switch. Proceed to the section in Chapter 21, "Mode switches."

VCR Stopping

The unit can be in Play, Fast-Forward, or Rewind modes. All three functions are performing properly. After 60 seconds or so, the unit stops.

FIGURE 20.4

Tape counter and its belt.

In older models, the cause of this problem is the counter belt slipping or coming off the groove on the take-up spindle, as shown in Figure 20.4. A sensor is incorporated in the tape counter. When the tape-counter pulley doesn't turn, this sensor signals the unit to stop.

You've inserted video tape and pushed Play. The unit runs for approximately 60 seconds with a good picture, then stops. Each time you push Play, you receive the same results.

The cause of this problem is that the take-up spindle isn't turning. Proceed to Chapter 23, "Take-up spindle problems."

You've inserted video tape and pushed Play. The unit might run for as long as 30 minutes before the unit stops. You push Play again; this time, the unit will play for 1 to 5 minutes and then stop.

The reason is that the take-up reel pulls in more video tape, which weighs down the reel inside the videocassette. The weight and drag of the video tape causes the take-up spindle to stop turning. Proceed to Chapter 23, "Take-up spindle problems."

VCR Freezes

You insert video tape and push Play. You can hear the VCR loading the video tape onto the tape path, but then the unit stops. When you push Play or Eject again, the loading motor runs, but nothing moves.

You remove the top cover and see that the roller guides won't finish loading the video tape and it won't Eject.

The unit is frozen. The problem could be that the cam gear assembly of the pinch roller has jumped a tooth and is out of alignment. Proceed to the section in Chapter 27, "Fourth type" under the subhead "Removal of a pinch roller bracket" and to Chapter 15, "Pinch-roller alignment." To reconfirm the problem, try to move the pinch-roller cam gears back and forth. If it wiggles just a little bit, this is not the problem. If it is solid as a rock, it is the problem.

You're watching a rented movie and suddenly the VCR starts speeding up and the unit stops playing. You try to push the Play button again, but the unit still won't work. You push the Eject button, but the videocassette doesn't eject. The unit is frozen.

The problem could be that the video tape is wrapped around the capstan shaft. Proceed to Chapter 27, "Fourth reason" under the subhead "Reasons why a VCR will eat a video tape."

VCR Goes Dead

Small children consider a VCR their piggy bank. They like to listen to coins drop through the air vents or the cassette door. Coins conduct electricity, and when they find their way to a circuit board, they short it out.

If a section of the VCR goes dead:

1. Unplug the unit.

2. Shake the VCR back and forth and listen for rattles.

3. Check all circuit boards for coins or any other foreign objects made of metal and remove them.

QUICK REPAIR

A favorite pastime for children is to stuff small toys and crayons through the cassette door. I've found items, such as pen caps and crayons that have jammed the unit. They have fallen into one of the roller guide tracks and stopped the unit from loading the video tape. A small toy might float around inside the unit and eventually find its way into the teeth of some gear and jam it. Remove the main top cover and go treasure hunting. You'll be amazed at what you can find in a child's VCR piggy bank.

4. Check for a blown fuse. In some cases, the foreign object might have shorted out a B+ line, causing one of the fuses to turn black inside. A B+ line is a source that feeds electricity to select sections in a VCR. Replace this fuse. (For replacing fuses, refer to Chapter 16, "Replacing fuses.")

VCR Won't Accept a Videocassette

You turned the power on. The cassette holder moves in about an inch, so you'll be unable to insert a videocassette. At the same time, the cassette holder moves in, the cassette light indicator comes on. The cassette holder will either stay in this position or it'll pop back and forth and either stop or shut down.

The problem is a bad tape sensor. Proceed to Chapter 21, "Tape sensors."

You insert a videocassette and it goes in part way, then back out. Or, the videocassette goes in all the way, part way out and back in and out a few times. Then, it will either load the video tape or eject the videocassette and, in some models, shut down.

The problem is a bad mode switch. Proceed to Chapter 21, "Mode switches."

Stuck in Fast Forward

You turn the power on and the unit goes directly into Fast Forward mode. The tape won't eject. Pushing all of the mode buttons has no effect.

Remove the main top cover, turn the power on, and view the take-up reel. If the take-up reel is spinning, but the supply reel isn't, the video tape has broken away from the supply spindle when rewinding. The tape wasn't properly secured to the reel. The tape sensors inside the unit signaled the unit to fast forward. To correct this problem, unplug the unit and plug it back in to reset the microcomputer. In some units, the video tape will automatically eject. If not, refer to the section in Chapter 26, "Manually removing video cassettes." After removing the video tape, the unit will start functioning properly. To repair the video tape, refer to Chapter 29, "Repairing a video tape."

Stuck in Pause

You've inserted video tape and pushed Play. When the picture appears on the screen, it seems to be in Pause mode. It'll stay in Pause for 60 seconds and then stop.

FIGURE 20.5

The pin going into the undercarriage that fell out.

The cause of this problem could be a bad tape drawer pin. A tape drawer pin holds the video tape away from the capstan shaft in fast forward or rewind. When in Play mode, the drawer pin moves out of the way, allowing the pinch roller to meet the capstan shaft. At the opposite end of the bracket, holding the tape drawer pin, is another pin that goes into the undercarriage. The bracket is made of plastic. If it cracks, the pin going to the undercarriage can fall out. The bracket and the pin are shown in Figure 20.5. If the pin is missing, the pinch roller can't go up against the capstan shaft; therefore, the capstan shaft can't pull the video tape through the video tape path. To correct this problem, locate the pin that is located inside the undercarriage. Insert it back into the bracket and glue it into place.

If you find the same symptoms, but your unit doesn't have a drawer pin, the cause could be the pinch-roller assembly. It's out of alignment or the spring mounted to the pinch-roller bracket is broken or has slipped off. Refer to Chapter 7, "Removing a pinch roller" and Chapter 27, "Removal of pinch roller brackets" and to the section in Chapter 15, "Pinch-roller alignment."

Another reason that a VCR gets stuck in pause is that the capstan shaft isn't turning and pulling the video tape through the tape path. Proceed to the section in Chapter 25, "Detecting a capstan shaft problem."

Clicking or Grinding Noise

You've inserted video tape and pushed Play. You hear a clicking sound coming from the unit.

In older models, the cause might be the tape counter. To correct this problem, place your fingernail on the edge of the reset button to the tape counter. Push the button in and let it pop off your fingernail. This will reset the gears in the tape counter.

In newer models, a small gear attached to the take-up spindle can cause the clicking noise. In Fast Forward mode, you might hear a grinding noise. The reels inside the videocassette aren't moving. The unit stays in this mode for about 60 seconds, then the unit stops. Proceed to the section in Chapter 15, "Take-up spindle alignment."

Loud Squeal

You've inserted video tape and pushed Play. You hear the unit load. Just before the picture should appear, you hear a loud squeal. The unit either unloads or shuts down.

Remove the main top cover, plug the unit in, and push Play. Watch the roller guides load the video tape to the tape path. Notice that the roller guides don't quite reach the V-mounts when you hear this loud squeal. The video tape isn't being pulled into the tape path, then the unit unloads or shuts down. The cause of this problem is a bad loading belt. First, refer to the sections in Chapter 19, "Loading motors" and "Removing loading motors and belts." After locating the loading motor, read the first four sections of Chapter 10. After locating the loading motor, plug the unit in and push Play. Look at the motor that's running and the belt that's squealing. This belt is the one you need to replace.

QUICK>>>TIP Countless times I've heard people say, "My VCR won't record anymore." If your VCR has this problem, look on the back right corner of the videocassette for a missing tab (Figure 20.6). This removable safety tab prevents the VCR from recording over previously recorded video tape. If this tab is missing, cover the hole with a piece of electrical or clear tape. You can now reinsert the videocassette and begin recording.

When playing, the unit has a good picture and sound, but you hear a squeal or screeching in the background coming from the unit. Sometimes the sound is present immediately after the unit starts to play or record, or it might appear 20 minutes later. Proceed to Chapter 22, "High-pitched sound problems," to the section in Chapter 19, "Checking for bad bearings" or to the section in Chapter 25, "Squeaking or squealing in Play mode."

Indicator Lights Flashing On and Off

A tape is inside of the unit and the indicator lights are flashing off and on. The function buttons are nonfunctional. Try unplugging the unit and plugging it back in. The function lights should stay on for about one minute, then they'll start flashing again.

The problem is a bad loading motor or a broken loading belt. First, refer to the sections in Chapter 19, "Loading motors" and "Removing loading motors and belts." After locating the loading motor, read the first four sections of Chapter 10.

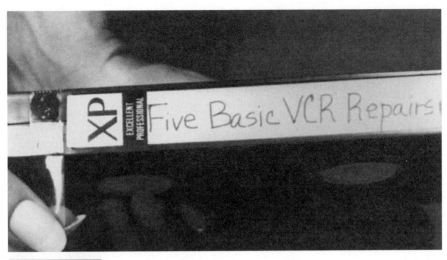

Placing a piece of tape over the missing tab.

Tape Sensor and Mode Switch Problems

Tape Sensors

VCRs have two tape sensors. The tape sensors serve three purposes:

1. To protect the videotape from breaking when it reaches the end of the reel when in Fast-Forward or Rewind mode.
2. To signal the unit that the video cassette is either at the beginning or at the end and to automatically stop and/or rewind the video tape.
3. To let the unit know that a videocassette has been inserted into the unit and the tape is ready to play.

Tape-sensor Functions

An infrared light-emitting diode (LED) protrudes through the transport in the middle of the cassette carriage. The two physical types are:

- A white or black plastic post with a bubble on each side near the top, as shown in Figure 21.1.
- The other type LED is located under the transport and the light is reflected up through and out the sides of the prism, mounted to the transport (Figure 21.2).

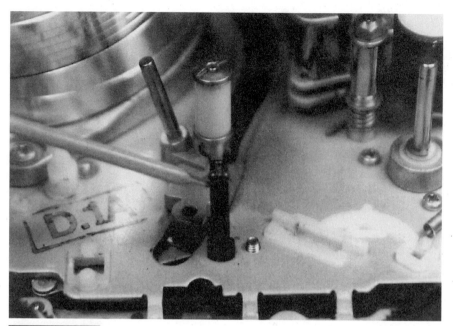

FIGURE 21.1

LED protruding up through the transport.

FIGURE 21.2

A prism mounted to the top of the transport above the LED.

The light from the LED or prism travels through the inside of the videocassette when it's inserted. On each end of a video tape is transparent leader tape. When the videotape reaches the end, the light shines through the leader tape, then through a small hole located on either end of the videocassette. The light then continues to travel to a tape sensor, which signals the unit to stop.

Tape-sensor Problems

The type of unit you have and whether or not the tape sensor shorts out or opens affects how your VCR will act. Read each reason and follow the instructions that fits your situation.

> **QUICK»TIP**
>
> Sometimes the outside room light will penetrate into the sensors after you remove the top cover and stop the unit from going into Play mode. It can also cause the unit to automatically go into the Fast-Forward or Rewind modes; this is disconcerting if you're adjusting or repairing the unit. In some cases, a VCR won't run at all because of too much room light. If this happens, remove or shade the light on the VCR.

FIRST REASON

You turned the power on, the cassette holder moves in about an inch so that you'll be unable to insert a videocassette. At the same time, the cassette holder moves in, the cassette light indicator comes on. The cassette holder will either stay in this position or it'll pop back and forth.

If you are unable to insert a videocassette, unplug the power and plug it back in to reset the microcomputer. This procedure returns the cassette holder to its loading position. In some models, this will correct the problem, but in most models, the problem will start up again in about five seconds. If so, replace both of the tape sensors.

If you insert a videocassette and it automatically ejects after two to five seconds or if the tape is pulled back into the unit after ejecting, shorted tape sensors are the cause of the problem. Before replacing the tape sensors, refer to the section in this chapter, "Mode switch problems." These types of problems can have similar symptoms.

SECOND REASON

You play a movie and the videotape goes all the way to the end where the transparent leader tape is connected. The light from the LED penetrates the end tape sensor.

If the sensor is working properly, the unit will go into Rewind automatically. In some models, after rewinding, the videocassette will eject. If the end sensor has opened, the unit will try to keep running and you will hear the video tape squeal at the pinch roller and capstan shaft for a few seconds. Then it will stop. The unit will not rewind.

THIRD REASON

You rewind videotape and when it reaches the transparent leader tape at the beginning of the tape, the light from the LED penetrates into the start tape sensor.

If the sensor is working properly, the unit will stop, then fast forward the leader tape back onto the take-up reel inside the videocassette, and stop again. If the start sensor is open, the unit will try to keep running and you will hear the video tape squeal at the capstan shaft, then stop, or the unit will break the videotape.

Locating Tape Sensors

The start and end tape sensors are located in a direct line from one another. One is on each side of the cassette carriage. The start-tape sensor is the sensor on the right side of the cassette carriage. The end-tape sensor is the sensor on the left side of the cassette carriage.

To locate the sensors, insert a videocassette. As the cassette holder drops down into its locked position, the front door on the videocassette opens up. Look along the top edge of this door, as shown in Figure 21.3. Follow this line to each end of the videocassette until you reach the cassette carriage. Then look straight down and you'll find a tape sensor.

Locate the cassette holder in the carriage. On each side of the cassette holder is a small round hole, approximately ⅛" to ¼" in diameter, as shown in Figure 21.4. When a videocassette is inserted and the cassette holder has dropped down into the carriage, these holes line up with the holes on the tape sensors. In some older models, the tape sensors are mounted on the sides of the cassette holder.

Identifying a Tape Sensor

Older models have the tape sensor mounted directly to the outside of the cassette carriage, one on each side, as shown in Figure 21.5. In other older models, a small circuit board is mounted to the outside of

FIGURE 21.3

Following the top edge of the cassette door to locate the tape sensors.

FIGURE 21.4

Tape sensor hole in the cassette holder.

FIGURE 21.5

A tape sensor mounted to the cassette carriage.

the cassette carriage, one on each side, as shown in Figure 21.6. The tape sensor is soldered to this board.

In newer models, the tape sensors are mounted to the mother board under the transport, as shown in Figure 21.7 (I removed the transport and drew arrows pointing at the sensors). The white or black plastic posts protrude up through the transport with the lens of the tape sensor located at the top of the post (refer to Figure 21.11). Some tape sensors slide up into a cover like the one shown in Figure 21.8 (I lifted the carriage up so that you can see the tape sensor under the cover). Most tape sensors have some kind of cover, but not all of them. Some cassette carriages have a prism mounted to the sides of the carriage above the tape sensors on the mother board, as shown in Figure 21.9. These prisms come in different sizes and shapes. Some units have a very small mirror mounted to the side of the cassette carriage above the tape sensors.

FIGURE 21.6

A tape sensor circuit board.

FIGURE 21.7

End tape sensor (left), LED (center), start tape sensor (right).

FIGURE 21.8

Cover for the tape sensor mounted to the carriage.

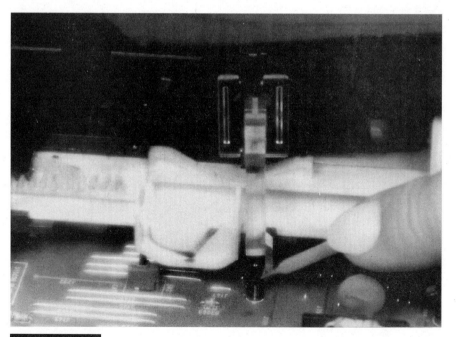

FIGURE 21.9

A prism mounted to the carriage above the tape sensor on the mother board.

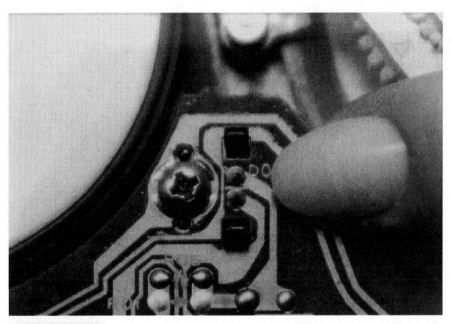

FIGURE 21.10

Two small black clips holding the mode switch in place.

Replacing a Tape Sensor

In older models with the tape sensors mounted to the cassette carriage, remove the carriage. Refer to Chapter 8, "Removing and servicing cassette carriages."

1. On each end of the carriage is a small circuit board (see Figure 21.6). At either end of this board is a small clip. Pull the clip away and pull the board out. (See Figure 21.10.)

2. Locate the tape sensor on the component side of the board.

3. Unsolder the two pins protruding through the board, as shown in Figure 21.10. Refer to the sections in Chapter 17, "Using solder wick to remove old solder" and "Soldering pins to a circuit board."

In other older models, you will find two little clips holding the tape sensors on the carriage (see Figure 21.5). Pull the clip away and pull the tape sensor off the carriage. In this type of unit, the wires at-

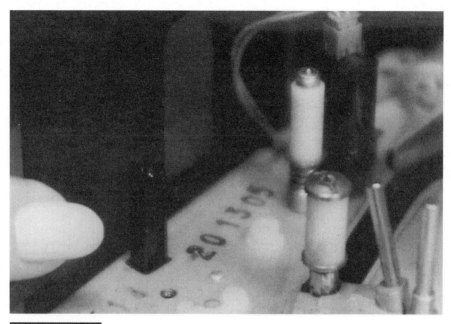

FIGURE 21.11

The bubble on top of the tape sensor post.

QUICK»TIP Each type of sensor has a little bubble on its side (see Figure 21.11). To insert the new sensor, be sure the bubble is facing toward the LED in the center of the cassette carriage.

tach directly to the pins of the tape sensors. Refer to the section in Chapter 19, "Soldering wires and terminals."

In newer models, the tape sensors are mounted to the mother board (refer to Figure 21.7).

1. Remove the transport. Refer to the section in Chapter 9, "Transports" and to the section in Chapter 31, "Removing the mother board." In other models, the sensors are mounted to a smaller circuit board under the transport.

2. To remove this board (refer to the section in Chapter 15, "Removing a circuit board mounted to the transport).

3. In either case, on the bottom of the board are two small black clips that protrude up through two rectangular holes, as shown in Figure 21.10. Unsolder the pins between the clips.

4. Refer to the section in Chapter 17, "Using solder wick to remove old solder" and "Soldering pins to a circuit board."

5. Squeeze the clips together and remove the tape sensor.

Mode Switch

The mode switch is connected to the moving mechanical parts in the VCR. The mechanical parts are those that are connected to the loading motor. The mode switch "tells" the microcomputer when all of the moving parts are in the proper position to play, record, pause, fast forward, rewind, and stop.

Mode Switch Problems

You insert a videocassette and it goes in part way, then back out. Or, the videocassette goes in all the way, part way out, or back in and out a few times, then either loads the videotape or ejects the videocassette and, in some models, shuts down.

You insert a videocassette and push Play. You can hear the VCR loading the videotape onto the tape path and before a picture is present on the TV monitor it unloads and then stops or shuts down. On the other hand, it might show a picture for a few seconds, it unloads and then stops or shuts down. You push Rewind or Fast Forward and the unit runs for a few seconds and then stops. Sometimes Fast Forward and Rewind will work okay and the problem is only in the play or record mode. The display on the front of the unit is showing the correct functions.

Removing Mode Switches

This section deals with locating and removing mode switches. Most mode switches are black in color and have many little holes punched in the back of them. They are round, oblong, or rectangular in shape (refer to Figures 21.16, 21.18, 21.22, and 21.28). They are usually located on top or under the main cam gear, on the loading motor housing bracket, or on a sliding arm or gear in the undercarriage. Read about each type of mode switch and follow the instructions that fit your situation.

FIRST TYPE

This mode switch is mounted on top of a circuit board in the undercarriage, as shown in Figure 21.12.

FIGURE 21.12

The mode switch mounted to the circuit board.

1. Remove the circuit board. Refer to the section in Chapter 15, "Removing circuits boards mounted to transports."

2. Flip it over and push in on each of the three tiny clips that hold the mode switch to the circuit board, as shown in Figure 21.13 (I drew arrows pointing at each clip). As you push in on a clip, simultaneously lift up on that portion of the mode switch.

3. Do each clip in the same manner until the top portion of the switch comes off. The part that you just removed from the board is called an *armature*. The armature holds the leaf spring brushes, as shown in Figure 21.14. The armature is attached to a shaft and rotates the brushes around inside the switch. On the circuit board side is a maze of circular-printed contacts that make contact with the brushes (see Figure 21.30).

4. Now that the switch has been taken apart, refer to the section in this chapter, "Repairing mode switches."

To remount the mode switch, align the three clips up to the holes on the circuit board and push the two pieces together until you hear each clip snap back together (see Figure 21.13). Under the circuit

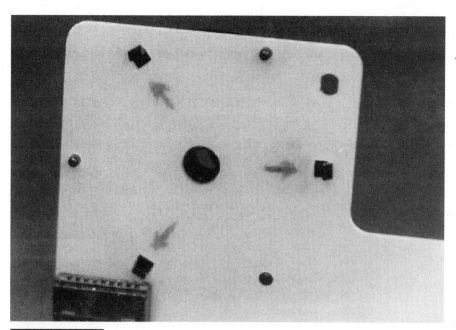

FIGURE 21.13

The three tiny clips holding the mode switch in place.

FIGURE 21.14

The leaf spring brushes on the armature.

FIGURE 21.15

The soldered pins from the mode switch.

board is a gear with a shaft protruding up. This type of shaft is called a *half-moon shaft*, meaning that one side of the shaft is flat (see Figure 21.23). The hole in the center of the mode switch has a half-moon hole (see Figure 21.13). Place a small screwdriver into the hole and turn the half moon until it aligns with the shaft. Now you can remount the circuit board to the shaft. This will align the mode switch to the gearing.

SECOND TYPE

This mode switch is mounted under a circuit board in the undercarriage and the pins to the switch are soldered to the circuit board, as shown in Figure 21.15.

1. Refer to the sections in Chapter 17, "Using solder wick to remove old solder" and "Soldering pins to a circuit board." Also, refer to the section in Chapter 15, "Removing circuit boards mounted to transports."

2. Remove the mounting screw and lift the mode switch straight up, as shown in Figure 21.16.

FIGURE 21.16

The mode switch and the mounting screw.

3. On the side of the switch is a white sliding shaft that connects to the switch inside. With a can of tuner cleaner, spray down into the slot where the shaft comes out, as shown in Figure 21.17.

4. Slide the shaft back and forth, from end to end. This action will clean the contacts inside the switch.

When remounting, be sure to place the shaft on the switch over the shaft on the sliding arm (refer to Figure 15.35). This will automatically align the mode switch to the gearing. Remount the mounting screw and the circuit board.

THIRD TYPE

This mode switch is mounted on top of the cam gear and the cam gear is on top of the transport.

1. Remove the two O-rings on top of the housing that partially covers the switch (refer to Figure 7.19).

2. Lift the housing straight up and off.

FIGURE 21.17

Spraying inside the mode switch with tuner spray.

3. Remove the two O-rings on top of the switch, as shown in Figure 21.18.

4. Unplug the white plug at the base of the switch and lift the switch straight up and off.

5. Take the mode switch apart.

6. Four latches are located around the outside of the switch. Place an ice pick under a latch and pry it up, as shown in Figure 21.19, while, at the same time, holding that portion of the switch apart.

7. Do each latch in the same way until the switch comes apart.

8. Refer to the section in this chapter, "Repairing mode switches."

9. After repairing the switch, place the two outside pieces so that all four latches are aligned and squeeze the two pieces together until the latches snap back together.

When remounting, be sure that the cam gear is aligned properly. Refer to the section in Chapter 15, "Main cam gear alignment." To re-align the mode switch, place the round alignment post on the armature

FIGURE 21.18

Two O-rings on top and the white plug to the mode switch.

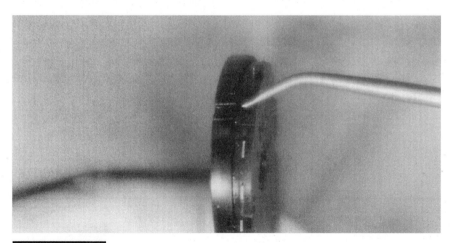

FIGURE 21.19

Using an ice-pick to pry the latch up.

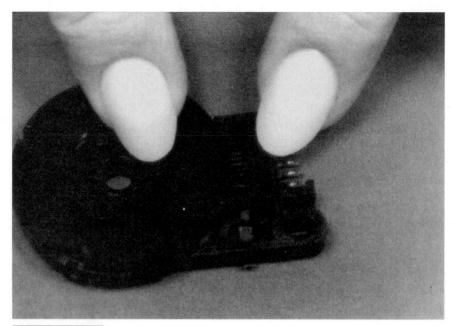

FIGURE 21.20

Aligning the post on the armature to the hole on the base of the mode switch.

(the part that moves in the center of the switch). Align the round post with the alignment hole on the base of the switch, as shown in Figure 21.20. Also on top of the armature are four square pegs, two on each side. A finger covers one of the square pegs, as shown in Figure 21.20. On the top of the cam gear, near the center, are two square pegs, just like on the armature. You need to place the two square pegs on the cam gear between each set of square pegs on the armature. Now that the cam gear and mode switch are aligned, slide the switch down the shafts where the O-rings were (see Figure 21.18). The two pegs on the cam gear will automatically slide up between each set of square pegs on the armature. This will align the switch to the gearing. Place the bracket back into place and remount the O-rings on top of the switch and bracket.

FOURTH TYPE

This mode switch is mounted to the transport in the undercarriage.

1. A clip runs through the transport. To remove the switch, push on the clip and pull the switch off the transport. Figure 21.21 shows the clip.

FIGURE 21.21

The clip (top) and the mounting hole (bottom) on the mode switch.

2. To take the switch apart, place a fingernail between the armature and the base of the switch.

3. On the back and near the center of the switch are four clips going to the armature, as shown in Figure 21.22. Use another finger to push in on each clip, one at a time, and simultaneously pull out on that section of the armature with the fingernail that is already placed between the armature and the base.

4. Do each clip until the switch comes apart.

5. Refer to the section in this chapter, "Repairing mode switches."

6. After repairing the switch, place the two pieces together and squeeze them until the clips on the armature snap back into place.

When remounting, a shaft protrudes up from the transport, as shown in Figure 21.23. This shaft is a half-moon shaft. The hole in the center of the mode switch has a half-moon hole. Turn the half-moon hole or armature until it aligns with the shaft on the transport. Slide the switch over the shaft and align up the clip on the switch to the

FIGURE 21.22

Placing fingernail between the armature and the base of the mode switch. The white clips to the armature.

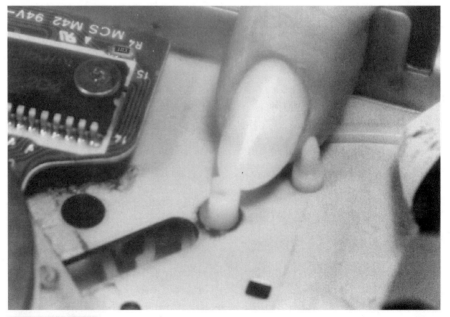

FIGURE 21.23

The half moon shaft.

The three mounting screws to housing bracket.

small rectangular hole in the transport and, at the same time, align the post on the transport to the hole on the switch (see Figure 21.21). Push the switch down until the clip snaps into place. This will align the switch to the gearing.

FIFTH TYPE

This mode switch is mounted to the housing bracket in the undercarriage.

1. Remove the three mounting screws to the housing, as shown in Figure 21.24 (I drew arrows pointing at the mounting screws).

2. Lift the housing straight up and off.

3. Flip the housing over and remove the two O-rings on the gears mounted to the housing, as shown in Figure 21.25.

4. Lift the two gears straight up and off.

5. Flip the housing over again.

FIGURE 21.25

The O-ring's on top of the gear's mounted to the housing bracket.

6. Five soldered pins are on the back of the mode switch, soldered to the ribbon (see Figure 21.26). Unsolder these pins. Refer to the sections in Chapter 17, "Using solder wick to remove old solder" and "Soldering pins to a circuit board."

7. Four clips are on top of the housing, as shown in Figure 21.26. To remove the switch, simultaneously lift up on that portion of the mode switch as you push in on a clip.

8. Do each clip the same way until you have removed the switch.

9. To take the switch apart and to reassemble, use the same method as shown in the "Fourth type."

To remount the mode switch, align the four clips up to the holes on the housing and push the two pieces together until you hear each clip snap back together (see Figure 21.26). Resolder the ribbon back onto the switch. Remount the large gear to the shaft protruding up from the center of the mode switch. Place the round pin on the armature, as shown in Figure 21.27, in the only hole near the center of the large gear above the switch. Turn the gear above the switch to align the hole

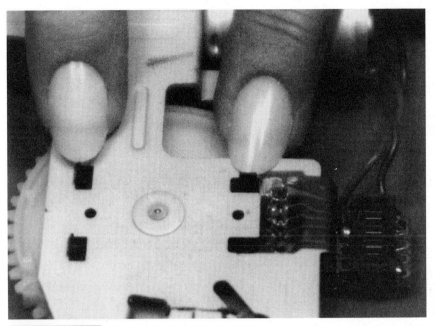

FIGURE 21.26

The four clips and ribbon to the mode switch on the back of the housing bracket.

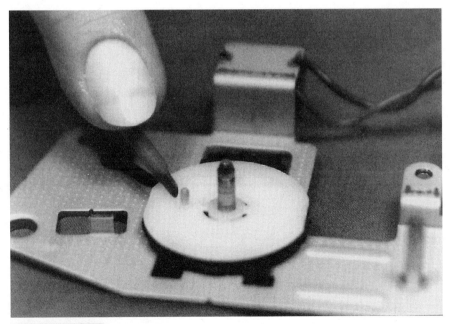

FIGURE 21.27

The alignment pin on the armature.

FIGURE 21.28

Mounting screw, mode switch, and lever.

in the gear to the hole in the housing bracket. Refer to the sections in Chapter 15, "Main cam gear alignment" and "Gear alignment." Remount the little gear and the O-rings of both gears (see Figure 21.25). Remount the housing to the transport and check for alignment to the gears on the transport. Refer to the sections in Chapter 15, "Main cam gear alignment" and "Sliding arm alignment."

SIXTH TYPE

This mode switch is mounted above a sliding arm in the undercarriage.

1. Remove the mounting screw and the lever that holds down the mode switch, as shown in Figure 21.28.

2. To remove the lever, you need to unlatch a latch at the top of the lever (Refer to Figure 15.41, this shows the same type of latch).

3. After you remove the lever, lift the switch straight up and off.

4. To take the switch apart and to reassemble, use the same method as shown in the "Fourth type."

FIGURE 21.29

Aligning the notch on the armature to the hole on the
mode switch.

When remounting, align the notch on the armature with the hole
on the mode switch, as shown in Figure 21.29. Check the sliding arm
alignment. Refer to the section in Chapter 15, "Sliding arm alignment." After you know that the sliding arm is aligned, slide the switch
down the shaft where the lever was mounted (see Figure 21.28). The
gear on the armature will automatically align with the teeth on the
sliding arm. Remount the mounting screw and the lever. Be sure that
the pin on the bottom of the lever is in the groove of the sliding arm
(refer to Figure 23.25).

Repairing Mode Switches

The problem with mode switches is that the contacts inside the switch
become dirty from dust in the air mixing with the grease inside the
switch, or by moisture in the air causing corrosion on the contacts.
The dirty or corroded contacts send mixed messages to the microcomputer and cause the unit to malfunction.

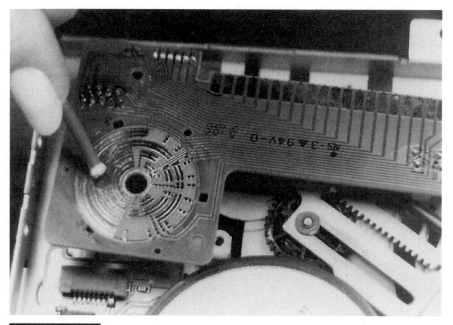

FIGURE 21.30

Cleaning the contacts on the circuit board.

1. Refer to the section in Chapter 1, "Making a glass brush." Dip a glass brush into some cleaning alcohol.

2. Use the glass brush to clean the contacts on the circuit board, as shown in Figure 21.30 or the contacts inside of the switch, as shown in Figure 21.31.

3. After cleaning each contact, re-clean the glass brush with a paper towel.

4. Spray the contacts with degreaser to remove any fibers left by the glass brush.

5. Clean the three to five leaf spring brushes by scrubbing each one at a time (see Figure 21.14).

6. Re-clean the glass brush with a paper towel after each leaf.

7. After cleaning each leaf, place the tip of an ice pick under one of the leaf springs. Pick up on the leaf so that the leaf will protrude up a little higher, as shown in Figure 21.32. You are bending the leaf up just a little so that the leaf will have more tension on the printed contacts, which makes for a better contact. Do each leaf in the same way.

FIGURE 21.31

Cleaning the contacts inside the mode switch.

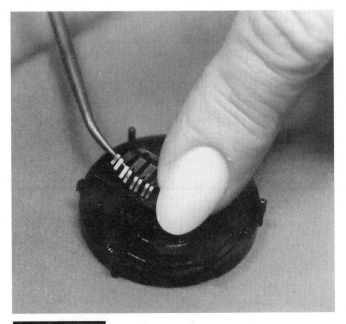

FIGURE 21.32

Using an ice-pick to bend the leaf up.

8. Place some tuner grease on the printed contacts and on each leaf.

9. Put the switch back together and/or remount the switch.

Review

1. The light from the LED travels up inside the videocassette. When the videotape reaches the end, the light shines through the leader tape into the sensors. This puts the unit into the Stop or Rewind modes.

2. A VCR won't run at all, or stays in Fast-Forward or Rewind modes because of too much room light. If this happens, remove or shade the light on the VCR.

3. The tape sensors are located in a direct line from one another. One is on each side of the cassette carriage.

4. The tape sensors can be mounted to the sides of the carriage or mounted to the mother board and protrude up through the transport.

5. The mode switch indicates to the microcomputer when all the moving parts are in the correct position to play, record, pause, fast forward, rewind, or stop.

6. The mode switch could malfunction, causing the unit to show a picture for only a few seconds, unload, and then stop or shut down.

7. Mode switches are located on top or under the main cam gear, on the loading motor housing bracket, or on a sliding arm or gear in the undercarriage.

8. Mode switches can be taken apart for repair.

9. Use the glass brush to clean the contacts inside of the mode switch.

High-Pitched Sound Problem

While viewing a movie, you suddenly hear a high pitched sound or a squealing coming from the VCR. The sound isn't coming from the speakers in the monitor. When the sound first occurs, it periodically comes and goes, but only in Play or Record modes. Later, the sound becomes fairly constant. When you stop the unit, the sound continues for another one to three minutes.

Locating the High-Pitched Sound

The sound is coming from a grounding leaf spring that is attached to the video drum. The grounding spring removes static electricity from the video drum. If the spring has been improperly cleaned or is dry, white dashes appear in the picture along with a high-pitched sound. There are three different locations at which you will find the grounding spring. Each location is covered in this section and has its own set of instructions. To find the location from which the sound is coming, read each of the following sections.

FIGURE 22.1

First location for a grounding leaf spring.

First Location

1. Remove the main top cover.

2. Refer to Chapter 2, "Getting inside the VCR."

3. Insert a videocassette and push Play.

4. As soon as you hear the high-pitched sound, look at the top of the video drum. You'll find the grounding leaf spring, as shown in Figure 22.1.

5. While the unit is producing this sound, push lightly on the spring. As soon as you touch the spring, the high-pitched sound or squeal should stop. Although the sound has stopped, this doesn't correct the problem. The sound is caused by a lack of lubrication.

6. In some models, the grounding leaf spring is inserted inside a metal bracket. You'll need to remove the grounding spring bracket by removing the Phillips head mounting screw at the base of the bracket, as shown in Figure 22.2.

FIGURE 22.2

Bracket holding the grounding leaf spring.

Second Location

In some models, the grounding leaf spring is in the undercarriage.

1. Place the unit on its side.

2. Refer to Chapter 9, "Getting into the undercarriage."

3. Place your finger on top of the video drum and spin it.

4. Look in the undercarriage for a small flywheel that's spinning. At the base of the flywheel is a leaf spring, as shown in Figure 22.3. To be sure that this is the problem, push Play. As soon as you hear the high-pitched sound, touch the spring.

Third Location

In other models, the grounding leaf spring isn't accessible. These types of units have a rotary cap on the top of the video drum. The cap must be removed to reach the spring.

1. Two mounting screws are located on top of the rotary cap, as shown in Figure 22.4.

Second location for a grounding leaf spring.

The two mounting screws for the rotary cap.

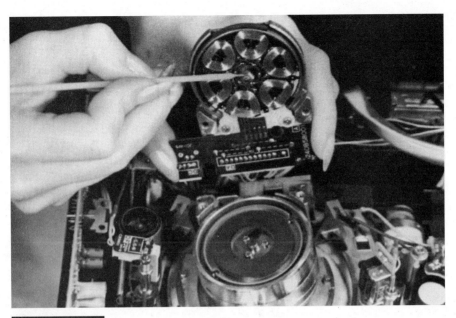

FIGURE 22.5

Third location for a grounding leaf spring.

2. Remove the screws and pick the rotary cap straight up and off. Magnets inside hold it down, so it'll be hard to pick up. After removing the cap, you'll see the leaf spring, as shown in Figure 22.5.

3. When remounting the rotary cap, you'll notice two small pins protruding from the bottom. On top of the mounting bracket, behind the video drum, are two pinholes beside each screw hole, as shown in Figure 22.6.

4. When replacing the rotary cap, line up the two pins with the pinholes. The magnets will cause the cap to clamp against the top of the video drum. Insert and tighten down the two mounting screws. As the screws are tightening down, the rotary cap will automatically align to the video drum.

Removing the High-Pitched Sound

1. Stop the unit and allow the video drum to stop spinning.

2. Saturate a chamois stick with cleaning alcohol.

FIGURE 22.6

A pin on the bottom of the rotary cap.

3. While slightly lifting up on the spring, slide the chamois stick between the bottom of the spring and the top of the upper video drum shaft.

4. Keeping a slight amount of pressure on the top of the spring, rub the chamois stick back and forth, as shown in Figure 22.7.

5. Reposition the chamois stick to remove the entire black residue.

6. Place some phono lube on the tip of a small flathead screwdriver and lift up the leaf spring.

7. Place the phono lube on top of the video drum shaft and the bottom of the spring, as shown in Figure 22.8.

In some models you need to remove the grounding spring.

1. Refer to the section in Chapter 1, "Making a glass brush."

2. Dip a glass brush into a small container of cleaning alcohol.

3. Use the brush to clean the bottom of the spring and the top of the upper video drum shaft.

4. Place some phono lube on top of the video drum shaft and reassemble.

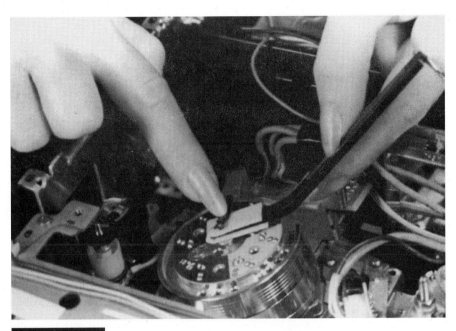

FIGURE 22.7

Cleaning the top of the video drum shaft and the bottom of the spring.

FIGURE 22.8

Lubricating the shaft and its leaf spring.

Take-Up Spindle Problems

The take-up spindle attaches to the take-up reel inside of the video-cassette when the cassette is inserted (see Figure 23.2). In the Play mode, the function of the take-up spindle is to pull the videotape back into the videocassette with the right amount of torque.

Take-Up Problems

You're viewing videotape while it plays. The picture and sound are good, but then the unit automatically unloads the tape back into the videocassette after running for only 30 to 60 seconds. Each time you push Play, the same thing happens.

1. Remove the main top cover.

2. Refer to Chapter 2, "Getting inside the VCR."

3. Insert the videotape and push Play.

FIGURE 23.1

The video tape bundling behind the pinch roller.

4. Look directly behind the pinch roller and capstan shaft. You'll see the video tape bundling up, as shown in Figure 23.1. This continues for approximately 30 to 60 seconds, then the unit automatically stops and pulls the tape back into the cassette (I've removed the top cover on the video cassette to give you a better view).

5. Push Play again. This time, watch the take-up reel inside the cassette, as shown in Figure 23.2. Notice if the take-up reel isn't turning, or is turning sporadically. You'll need to look through the window on top of the cassette to see the take-up reel.

The unit might take as long as 30 minutes before this problem exists. The reason is that the take-up reel pulls in more tape, which weighs down the reel. The weight and drag of the videotape causes the take-up spindle to stop turning.

FIGURE 23.2

The take-up reel.

Take-up Spindle Systems

Several types of take-up spindle systems are used in VCRs. All take-up systems have one thing in common, an idler wheel. All idler wheels have a swivel point. When the unit is in the Play mode, the idler wheel will slide over and make contact with the take-up spindle. The drive wheel on top of the clutch assembly drives the idler wheel. The clutch in the undercarriage is connected to the drive wheel. A belt drives the clutch and the capstan motor drives the belt. The spindle, idler wheel, and the drive wheel can have a build-up of residue. This residue causes the wheels to be slippery or cause the clutch and belt to the clutch to slip and cause the VCR to malfunction. Read the following sections for the location of each type of system and how to correct each problem.

First Type

In older models, the system was a combination of a drive wheel, an idler wheel, and a take-up spindle. The idler wheel is to the right of the capstan shaft and behind the take-up spindle, as shown in Figure 23.3.

The take-up idler wheel.

The drive wheel is at the base of the capstan shaft, as shown in Figure 23.4. Drive wheels of this type barely stick out from under the base. When you push Play, the idler wheel moves over and makes contact simultaneously with the drive wheel at the base of the capstan shaft and the take-up spindle. To correct a take-up problem, refer to the section in this chapter, "Checking for slippage."

Second Type

The second type of system drives the take-up, fast-forward, and rewind. An idler arm assembly uses a rubber tire between the spindles (refer to Figure 24.19). When you push Play, the idler arm assembly moves over and makes contact with the base of the take-up spindle. To check for take-up problems, refer to the section in this chapter, "Checking for slippage." Because this system relates to fast forward and rewind, proceed to Chapter 24, "Fast forward and rewind problems."

Third Type

Newer models have no rubber tire on the idler wheel. Instead, you'll find an idler arm assembly with an idler gear, as shown in Figure 23.5.

FIGURE 23.4

The take-up drive wheel.

FIGURE 23.5

Idler arm assembly using a gear.

FIGURE 23.6

Two idler gears between the spindles.

On the other hand, you will find two idler gears between the spindles, as shown in Figure 23.6. When you push Play, an idler gear will move over to the teeth on the take-up spindle and drive it. The idler gear is attached to the drive gear, as shown in Figure 23.7. The drive gear is attached to the clutch assembly. The entire mechanism uses gears, with the exception of a belt in the undercarriage; there is no rubber or plastic to get dirty or cause slippage. Check the undercarriage for a bad belt and refer to the first four sections of Chapter 10. Then, check for a bad clutch assembly. Refer to the section in this chapter, "Checking clutch assemblies."

Fourth Type

The fourth type uses a direct-drive system. You'll find no idler or drive wheel on top of the transport between the spindles (refer to Figure 19.13). To check this type of system, refer to the section in Chapter 19, "Reel motors." In some direct-drive units, a small white gear is beside the take-up spindle. Sometimes this gear will slip up or down the shaft, causing the spindle to stop turning. Refer to the section in Chapter 15, "Take-up spindle alignment."

FIGURE 23.7

The point of contact of the idler gear and drive gear on top of the clutch assembly.

Checking for Slippage

This section deals with VCRs that use a rubber tire.

1. Insert a blank cartridge into the VCR (refer to Figure 1.1).

2. Push Play and hold the take-up spindle to keep it from turning, as shown in Figure 23.8.

3. While holding the take-up spindle to keep it from turning, the unit will stop periodically. There's nothing wrong. A sensor signals the unit to stop when the take-up spindle isn't turning.

4. When the sensor stops the unit, push Play again.

In models where the idler wheel stops turning, but the drive wheel at the base of the capstan shaft is turning, the slippage is between the idler wheel and the drive wheel, as shown in Figure 23.9. On the other hand, if the idler wheel is turning, then the slippage is between the idler wheel and the take-up spindle (refer to Figure 24.3). If both wheels stop turning, a bad belt or a bad clutch in the undercarriage

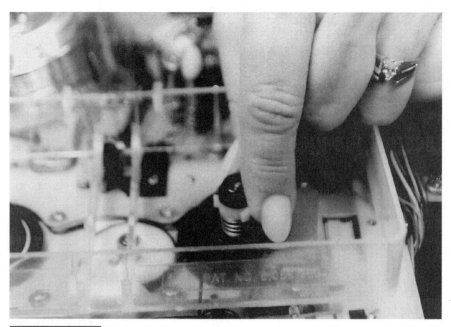

FIGURE 23.8

Stopping the take-up spindle from turning.

FIGURE 23.9

The point of contact for the take-up idler wheel and the drive wheel.

causes the problem. Refer to the section in this chapter, "Checking clutch assemblies."

In older models, all idler wheels are surrounded by a rubber tire. The drive wheel and most spindles are made of plastic. To clean the idler wheel, refer to the section in this chapter, "Cleaning rubber tires." To clean the spindle, refer to the section in this chapter, "Cleaning wheels made of plastic." To clean the drive wheel at the base of the capstan shaft or the drive wheel on top of the clutch assembly, refer to the section in this chapter, "Cleaning the drive wheel." If the unit has no rubber tires and uses gears instead of wheels, refer to the section in this chapter, "Checking clutch assemblies."

Cleaning Rubber Tires

1. Fold a piece of #400 to #600 (fine grain) sandpaper in half.
2. Sand the tire from side to side, as you slowly turn the wheel, as shown in Figure 23.10.
3. Sand all the way around the tire.
4. If the tire is shiny, has cracks, or has a notch in the rubber, you must replace the tire. Refer to the section in Chapter 24, "Replacing tires."

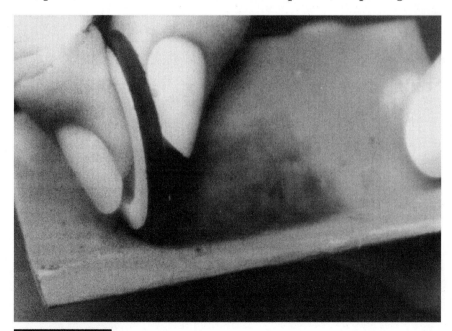

FIGURE 23.10

Sanding a rubber tire.

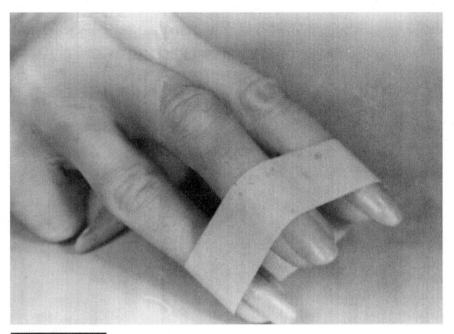

FIGURE 23.11

Making a ring from masking tape with the sticky side out.

To clean rubber tires in the unit:

1. Cut a 6-inch piece of masking tape and make a ring out of it with the sticky side out.

2. Place your hand inside the ring, as shown in Figure 23.11.

3. Place your thumb and index fingers on each side in the center of the wheel and place the masking tape up against the rubber on the tire, as shown in Figure 23.12.

4. Watch the masking tape as you rotate the wheel. As the black residue builds up on the tape, reposition it to a clean spot and continue until the tape stops showing black residue.

QUICK»»TIP After cleaning, recheck the wheel for white spots left on the rubber tire. Sometimes the white adhesive from the masking tape comes off, causing white spots. While holding the wheel still, stick a new piece of masking tape over each white spot and pull the tape back off. You might have to repeat this process several times to remove the spots.

FIGURE 23.12

Cleaning off the black residue from the tire.

To clean rubber tires on wheels that have been removed from the unit:

1. Place a piece of masking tape with the sticky side up on a table.
2. Hold one end down with a finger and roll the wheel over a piece of masking tape, as shown in Figure 23.13.
3. Watch the masking tape as you roll the wheel. As the black residue builds up on the tape, reposition it to a clean spot.

Cleaning Wheels and Spindles Made of Plastic

To clean wheels made of plastic, saturate a chamois stick with head cleaner or alcohol. Rub the chamois stick in an up and down motion as you go around the wheel or spindle, as shown in Figure 23.14. If the wheel still slips after using a chamois stick, then use masking tape to remove the residue as previously explained. The exception is that you dab the wheel with the tape as you work around it. Check for white

Rolling a rubber tire on a wheel over masking tape.

Cleaning a plastic spindle with a chamois stick.

spots after using masking tape and remove them as previously explained.

Cleaning the Drive Wheel

1. With a can of degreaser, spray off the drive wheel (refer to Figures 23.4 and 24.2).

2. Use a glass brush and remove any black residue left on the drive wheel and re-spray it with degreaser.

3. To rotate the wheel to clean the other side, place your finger on top of the flywheel on top the capstan motor and spin it in either direction (refer to Figure 4.7). This motion will cause the drive wheel to turn.

To turn the drive wheel in all other models, refer to Chapter 9, "Getting into the undercarriage" and turn the large flywheel at the base of the capstan shaft (refer to Figure 10.2).

Checking the Tire

Sometimes after cleaning the take-up spindle, the tire on the idler wheel, and the drive wheel, the take-up spindle still doesn't turn properly. If this situation occurs, you'll need to replace the tire on the idler wheel. First, refer to the section in Chapter 24, "Replacing tires." If you need to remove the idler wheel assembly to replace the tire, refer to the sections in Chapter 24, "Removing idler wheels" and "Removing idler wheels and clutch assemblies," or to the section in this chapter, "Removing a take-up idler wheel." If the idler wheel doesn't make good contact against the take-up spindle while in the Play mode, the clutch assembly is slipping and the clutch assembly might need to be replaced. Refer to the section in this chapter, "Checking clutch assemblies."

Removing a Take-up Idler Wheel

In older models, to remove the idler wheel assembly and replace the tire:

1. Remove the cassette carriage (refer to Chapter 8, "Removing and servicing cassette carriages").

2. Remove the two brackets, as shown in Figure 23.15, by removing the Phillips head screws at the base of each bracket.

FIGURE 23.15

Removing the two brackets.

3. Remove the C- or E-ring above the pinch-roller assembly, as shown in Figure 23.16.

4. To remove C- or E-rings, refer to the section in Chapter 7, "Removing C- and E-rings."

5. Lift the pinch roller and pinch-roller bracket straight up and off.

6. Remove the Phillips head screw on the left, as shown in Figure 23.17, and lift the bracket above the idler wheel straight up and off.

7. Remove the coil spring attached to the idler wheel bracket, just above the idler wheel.

8. Remove the C- or E-ring on the right, as shown in Figure 23.17, and lift the idler wheel and its bracket straight up and off.

9. Refer to the section in Chapter 24, "Replacing tires."

To remount the idler wheel, reverse the process. A few older models have similar take-up idler wheel assemblies. This will give you a good idea on how they should come apart.

FIGURE 23.16

The C-ring above the pinch roller bracket.

FIGURE 23.17

Removing the screw on the left and the C-ring on the right.

Checking Clutch Assemblies

The clutch is used for take-up speed only. The two plastic plates have felt between them (see Figure 23.19). One plate is fastened to the pulley where the belt is attached. The other plate is fastened to the drive wheel or drive gear, which drives the idler wheel or idler gear. One of these plates has a heavy-duty coil spring applying pressure across the felt onto the other plate, causing the right amount of slippage to the take-up spindle. The reason for the slippage is that the take-up reel will spin faster than the capstan shaft. This would cause the videotape to speed up and be eaten or the VCR would run at the wrong speed.

In older models, insert a blank cartridge and place the VCR on its side (refer to Chapter 9, "Getting into the undercarriage"). Locate the belt and pulley on the clutch assembly (refer to Figure 10.2, the clutch pulley is on the left). Push Play and manually stop the take-up spindle from turning. If the pulley under the clutch is turning and the drive wheel is not turning and you cannot feel any, or very little, torque on the spindle, then the clutch inside the assembly is worn or broken. If the pulley under the clutch isn't turning, the belt is worn and is slipping.

In newer models, the transport has to be removed to reach the clutch assembly. Insert a blank cartridge, push Play and manually stop the take-up spindle from turning. Check the drive gear to see if it's turning (see Figure 23.7). If it isn't and the capstan shaft is turning, and you cannot feel any (or very little) torque on the spindle, the problem is a bad belt or clutch. Remove the transport and read the first four sections in Chapter 10. If that checks out okay, then replace the clutch assembly.

Insert a blank cartridge and push Play. The idler wheel doesn't come completely into contact with the proper spindle. The idler wheel is spinning, but the spindle isn't turning. This condition means that the clutch attached to the drive wheel is bad. You'll need to replace the clutch assembly.

Another problem is when you push Play and the idler arm won't flop over to its proper spindle. The idler arm doesn't move and the idler wheel doesn't turn. The belt might be broken. Refer to the first four sections in Chapter 10. For replacement of the belt or clutch assembly, refer to the section in Chapter 1, "Parts."

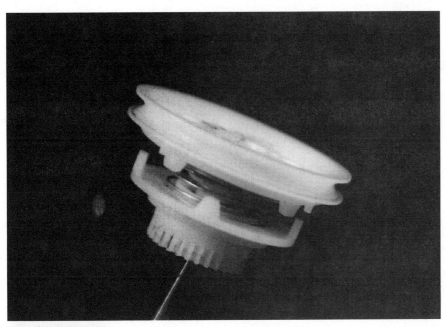

The pulley, coil spring, and drive gear on the clutch assembly.

Removing Clutch Assemblies

In a very few models, you might need to remove the cassette carriage. Refer to the section in Chapter 8, "Removing and servicing cassette carriages." To remove the transport, refer to the section in Chapter 9, "Transports."

Several types of clutch systems are used in VCRs. First, you need to be able to recognize what a clutch looks like. All clutches look similar and are made of plastic. In most cases, the drive belt from the flywheel is attached to the pulley on the clutch. A coil spring is in the center, between the pulley and the plate, above the drive wheel or gear, as shown in Figure 23.18. A flat pressure plate is on top of the pulley; it is attached through the center of the clutch to the drive gear. The pad is between the pressure plate and the top of the pulley.

First Type

In this clutch, you can see the felt pad sandwiched between two gears, as shown in Figure 23.19. The coil spring is inside the clutch assem-

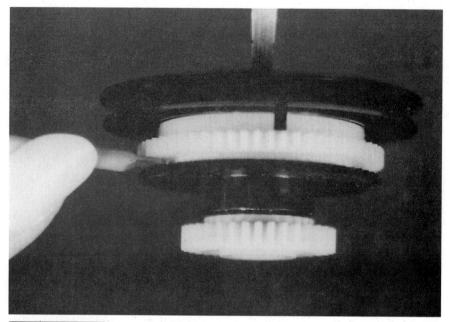

FIGURE 23.19

Pointing at the felt pad sandwiched between two gears.

bly, between the lower gear and the small drive gear, which drives idler wheel.

Second Type

In this clutch, you can see the felt pad sandwiched between two gears and the coil spring is on top, as shown in Figure 23.20. No pulley is attached to this clutch. To recognize the clutch, look for a felt pad sandwiched between two round plates or gears and a coil spring between the plates or a spring on the outside of one of the gears.

Third Type

In most models, a belt is attached to the clutch assembly. Remove the belt (refer to Figure 10.2). Remove the O-ring in the center of the pulley, as shown in Figure 23.21, and refer to the section in Chapter 7, "Removing O-rings." Lift the entire clutch assembly up and off its shaft. A small washer is at the bottom of the shaft; sometimes this washer sticks to the bottom of the clutch, don't lose this washer.

FIGURE 23.20

The coil spring on top of the clutch assembly.

FIGURE 23.21

The O-ring at the center of the pulley of the clutch assembly.

FIGURE 23.22

Pointing at the clutch, idler gear to the right, pulley and drive gear behind the idler gear.

Fourth Type

In some models, only the pulley will come off the shaft. This is because the clutch is beside the pulley (see Figure 23-20). Remove the belt and the O-ring in the center of the pulley. Lift the pulley straight up and off. Next, remove the O-ring in the center of the clutch and lift the clutch straight up and off.

Fifth Type

A few models are a little more complicated.

1. A gear is attached to the bottom of the pulley. The gear under the pulley is connected to the idler gear. The idler gear is connected to the bottom teeth of the clutch, as shown in Figure 23.22.

2. Remove the belt and the O-ring in the center of the pulley.

3. Remove the pulley, which has a drive gear attached to its bottom.

4. Remove the O-ring and lift off the washer, the coil spring, and the idler arm assembly (Figure 23.23), which was under the pulley.

FIGURE 23.23

The O-ring, coil spring, and idler arm assembly.

5. The ejection gear to the right of the clutch is hanging over the clutch. It must be removed before removing the clutch, as shown in Figure 23.24.

6. Remove the lever above the ejection gear by pulling out on the latch at the top of the lever at the shaft (refer to Figure 15.41; this shows the same type of latch).

7. Use an ice pick to pry the latch out; at the same time, lift up on the lever for removal.

8. Remove the O-ring at the center of the ejection gear and remove the gear.

9. Last, remove the O-ring on top of the clutch and remove the clutch (see Figure 23.22).

To reassemble, just reverse the process. No gear alignment is involved, but a pin protrudes out from under the lever. The pin on the lever needs to be placed back into the groove of the sliding arm, as shown in Figure 23.25.

FIGURE 23.24

The latch on top of the lever and the O-ring in the center of the ejection gear.

FIGURE 23.25

Where the lever pin goes into the groove of the sliding arm.

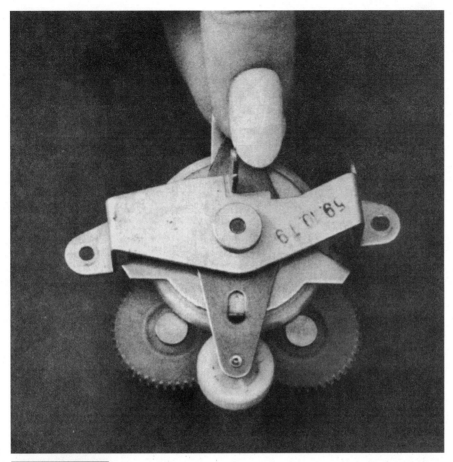

The clutch assembly, the idler arm and two mounting screw holes.

Sixth Type

In other models:

1. Two idler gears are on top of the transport (see Figure 23.6). Remove the two screws on the bracket above the two gears.

2. Lift the bracket straight up and off.

3. Lift the idler arm (with the two gears attached to it) straight up and off. Under the idler arm assembly is a clutch assembly.

4. Remove the O-ring above the clutch and lift the clutch straight up and off.

5. In some models, you need to get into the undercarriage and remove the belt from the pulley and/or the clutch assembly.

6. Flip the unit over and remove the Phillips head screws, one on each side of the clutch assembly, as shown in Figure 23.26.

7. Lift the entire assembly straight up and out (I removed the clutch assembly for a better view).

There are different ways to remove the clutch assemblies, but most of them are easy. Refer to the section in Chapter 24, "Removing idler wheels and clutch assemblies."

Review

1. If the VCR stops after 60 seconds while in the Play mode, you might have a take-up spindle problem.

2. Check to see if any excess tape is coming out from behind the pinch roller and capstan shaft.

3. Check to see if the take-up spindle and all connecting wheels are turning. If they are not, check the undercarriage for a bad belt or clutch.

4. Check for slippage between each wheel and clean them if needed.

5. If a rubber tire is shiny, has cracks, or a notch in it, replace the tire.

6. If the idler wheel doesn't make contact with the spindle, check for a bad clutch assembly and replace it.

7. To recognize the clutch, look for a felt pad sandwiched between two round plates or gears and a coil spring between the plates or a spring on the outside of one of the gears.

8. If you have a direct-drive system, check the DC motor.

Fast-Forward and Rewind Problems

Three problems commonly occur in Fast-Forward and Rewind modes:

■ Either Fast Forward or Rewind moves the video tape very slowly from reel to reel in the cassette.

■ Rewind stops before the entire video tape is back onto the take-up reel.

■ Both Fast Forward and Rewind stop immediately after you push the proper mode button.

Fast-Forward System and How It Functions

Starting in the undercarriage, the capstan motor drives a belt that is attached to a clutch assembly. The clutch assembly is attached to the drive wheel. The drive wheel protrudes up through a hole in the transport and drives the idler wheel. The idler wheel flops over to the take-up spindle and drives the spindle. The spindle drives the reel inside the videocassette.

FIGURE 24.1

The idler arm assembly.

There are three types of fast-forward systems. Read the following sections and follow the instructions that pertain to the model you have.

First Type

Older models have an idler arm assembly that uses a rubber tire between the spindles, as shown in Figure 24.1. The idler wheel is attached to the drive wheel, as shown in Figure 24.2. The drive wheel is

FIGURE 24.2

The drive wheel on the clutch assembly.

attached to the clutch assembly or the reel motor in the undercarriage. When you push Fast Forward, the idler arm assembly moves over and makes contact with the base of the take-up spindle and drives it. Refer to the section in this chapter, "Checking the fast forward system."

Second Type

Newer models have no rubber tire on the idler wheel. Instead, you'll find an idler arm assembly with an idler gear on it (refer to Figure 23.5). On the other hand, two idler gears are between the spindles (refer to Figure 23.6). When it fast forwards, the idler gear will move over to the teeth on the take-up spindle and drive it. The idler gear is attached to the drive gear (refer to Figure 23.7). The drive gear is attached to the clutch assembly. The entire mechanism uses gears, with the exception of a belt in the undercarriage; there is no rubber or plastic to get dirty or cause slippage. Check the undercarriage for a bad belt. Refer to the first four sections of Chapter 10. Check for a bad clutch assembly. Refer to the section in this chapter, "Clutch assemblies."

Third Type

The third type of system uses a direct-drive system. No idler or drive wheel is on top of the transport between the spindles (refer to Figure 19.13). To check this type of system, refer to the sections in Chapter 19, "Reel motors," "Checking motors for dead spots," and "Checking for dead motors."

Checking the Fast-Forward System

This section deals with VCRs that use a rubber tire in the fast-forward system.

1. Insert a blank cartridge (refer to Figure 1.1) and push Fast-Forward.

2. Hold the take-up spindle to stop it from turning (refer to Figure 23.8).

3. While holding the take-up spindle, the unit will stop periodically and unload. Nothing is wrong. A sensor signals the unit to stop when the take-up spindle isn't turning.

4. When the sensor stops the unit, push Fast-Forward again.

QUICK>>>TIP If the idler wheel is turning, but doesn't make complete contact with either spindle, the belt attached to the clutch assembly is slipping or the spring attached to the idler wheel is missing.

5. While holding the spindle, if both the drive wheel and the idler wheel continue turning while you're holding the take-up spindle, the problem is between the idler wheel and the take-up spindle, as shown in Figure 24.3. If the idler wheel stops turning, then the slippage is between the idler and drive wheel (see Figure 24.9). Clean all three wheels.

6. Refer to the sections in Chapter 23, on "Cleaning rubber tires" and "Cleaning wheels and spindles made of plastic." If the cleaning didn't repair the problem, proceed to the section in this chapter on "Replacing rubber tires."

If all the wheels have stopped, the problem is a bad belt or a bad clutch assembly in the undercarriage. Refer to Chapter 9, "Getting into the undercarriage." To clean belts and pulleys, read the first four sections in Chapter 10. When checking for a bad clutch assembly, refer to the section in this chapter, "Clutch assemblies."

FIGURE 24.3

Point where the idler wheel makes contact with the take-up spindle.

The point where the idler wheel makes contact with the supply spindle.

The Rewind System and How It Functions

When you push Rewind, the capstan motor runs in the opposite direction causing the idler wheel or idler gear assembly to flop over to the supply spindle, as shown in Figures 24.4 and 24.5. The only difference between the fast forward and the rewind system is that the supply spindle is connected to the system. All the same parts used in fast forward are also used in rewind, except for the take-up spindle.

To check the rewind systems using a rubber tire, insert a blank cartridge and push Rewind. Hold the supply spindle to stop it from turning. Use the same method as in this chapter, "Checking the fast-forward system." The unit will periodically stop as previously explained. Push Rewind again.

To check the systems using an idler gear and drive gear, check the undercarriage for a bad belt. Refer to the first four sections of Chapter 10. To check for a bad clutch assembly, refer to the section in this chapter, "Clutch assemblies."

FIGURE 24.5

The point where the idler gear makes contact with the teeth on the supply spindle.

The Idler Wheel

In older models, the majority of the time, the tire on the idler wheel causes the slippage. If the cassette carriage is in the way, refer to the section in Chapter 8, "Removing and servicing cassette carriages." First, clean each spindle and the drive wheel. Refer to the section in Chapter 23, "Cleaning wheels and spindles made of plastic." Then, try cleaning the idler wheel. Refer to the section in Chapter 23, "Cleaning rubber tires." If that doesn't work, replace the tire on the idler wheel. Refer to the sections in this chapter on "Removing idler wheels," "Removing idler wheels and clutch assemblies," and "Replacing tires."

In some models, the top of the clutch assembly covers most of the idler wheel.

1. Check this system, as previously explained, except notice that the idler wheel is hard to reach.

2. To clean the idler wheel, place a small flathead screwdriver on one side of the wheel to steady it.

3. Use a glass brush to clean the other side, as shown in Figure 24.6.

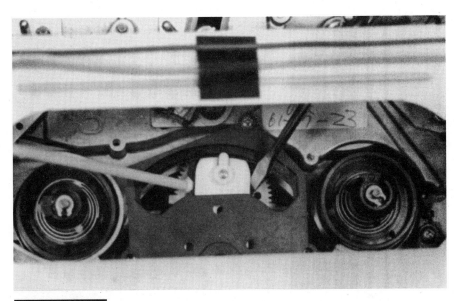

FIGURE 24.6

Using a glass brush to clean the idler wheel.

4. Use the screwdriver to rotate the wheel. Now, you can clean the next section of the wheel.

5. After cleaning the idler wheel, if the unit still doesn't work correctly, then you'll need to replace the tire on the idler wheel. Refer to the section in this chapter, "Removing idler wheels and clutch assemblies" and "Replacing tires."

In another type of system, a shaft sticks through a curved slot on the transport, as shown in Figure 24.7. The shaft is attached to the idler wheel. Check this system for slippage, as previously explained. In this type of unit, replace the tire on the idler wheel. Refer to Chapter 9, "Getting into the undercarriage." Refer to the section in this chapter, "Removing idler wheels and clutch assemblies" and "Replacing tires."

Another type of system has a large idler wheel between both spindles, as shown in Figure 24.8. No idler arm is attached to the idler wheel. In Fast-Forward or Rewind modes, the idler wheel itself slides over to the proper spindle. The drive wheel that drives the idler wheel is in front of the idler wheel under the metal bracket, as shown in Figure 24.9. Check this system for slippage, as previously explained. Clean the idler wheel. Refer to the section in Chapter 23, "Cleaning

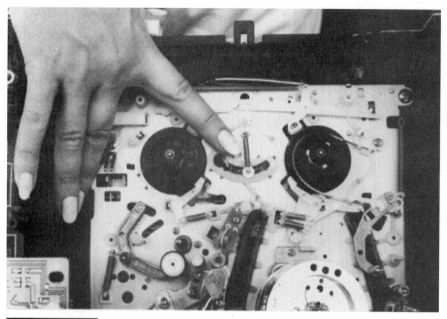

FIGURE 24.7

An idler wheel shaft protruding through a curved slot between the spindles.

FIGURE 24.8

An idler clutch assembly with no arm attached.

The drive wheel connecting to the idler wheel.

rubber tires." If that doesn't work, replace the tire. Refer to the sections in this chapter, "Removing idler wheels" and "Replacing tires."

Replacing Tires

To remove the tire, insert a cuticle remover and nail cleaner tool or a nut digger between the plastic groove on the idler wheel and the tire, as shown in Figure 24.10. Pry the tire out and be cautious not to break the plastic flange on the side of the groove. Stretch the tire straight out and fold it back over the wheel, as shown in Figure 24.11. To purchase a new tire, refer to the section in Chapter 1, "Parts."

To remount the tire, place one side of the tire in the groove. Stretch the other side over the wheel and into the groove. To replace any other tires, such as on spindles or drive wheels, use the same procedure.

Sometimes when you pop the tire onto the idler wheel, it won't seat correctly into the groove, making the tire look twisted on the idler wheel. With the nail tool, push down on the side of the tire where it is twisted, as shown in Figure 24.12. This step will place the tire com-

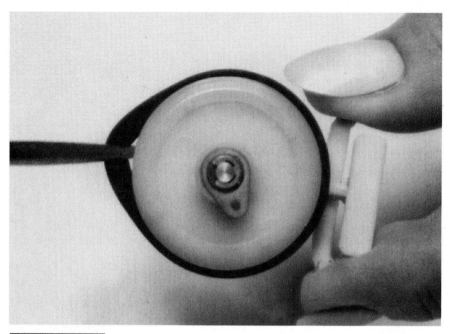

FIGURE 24.10

Insert the tool between the wheel and tire.

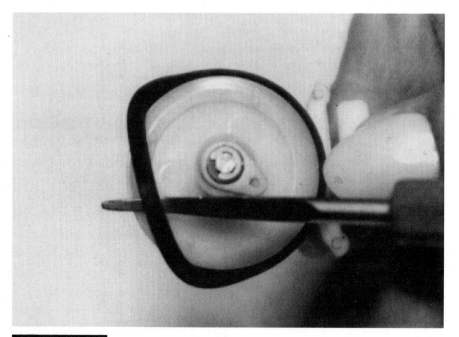

FIGURE 24.11

Folding the tire back over the wheel.

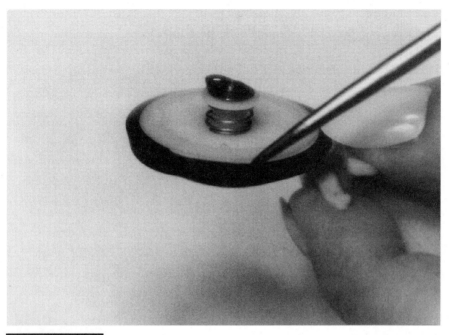

Using a tool to push the twisted tire down into the groove.

pletely and correctly into the groove. Another difficulty is that the tire isn't all the way down into the groove, making the tire look lopsided. With your fingers, squeeze the outside of the tire until the tire is even all the way around.

Clutch Assemblies

Read the section in Chapter 23, "Checking clutch assemblies" and the first three sections under "Removing clutch assemblies." These sections explain what a clutch assembly looks like and how it works.

When you push either the Fast Forward or Rewind buttons, the loading motor starts up and turns the main cam gear. There are levers, sliding brackets, and gears attached to the cam gear. Some of these moving parts are attached to the clutch assembly. When these parts are positioned just right in Fast Forward or Rewind, the mode switch tells the loading motor to stop. At the same time, the clutch is disconnected or bypassed by these moving parts. The clutch is bypassed so that the spindles are directly connected to the capstan motor. Then, the spindles can speed up and have no slippage.

The clutch can be disconnected in three different ways. Read the following sections and choose the one that pertains to the model you have.

First Way

A solid double sandwich gear is beside the clutch. The loading motor moves the solid sandwich gear over to the clutch gears and attaches the gears on the solid sandwich gear to the clutch, as shown in Figure 24.13. Now the clutch is disconnected and you have direct drive to the spindles.

Second Way

At the base or under the clutch is an arm, as shown in Figure 24.14 (I removed the clutch to give you a better view of the arm). The loading motor moves the arm up and pushes the clutch plates together. On the inside of each plate, where the coil spring is located, teeth protrude to-

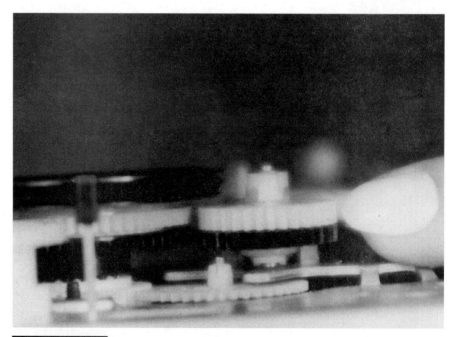

FIGURE 24.13

A solid double sandwich gear (right), a double sandwich gear on the clutch (left).

FIGURE 24.14

The arm at the base of the clutch.

ward each other (refer to Figure 23.18). The teeth come together and bypass the clutch. Now you have direct drive to the spindles.

Third Way

The idler gear is right beside the clutch, in the undercarriage attached to the bottom gear on the clutch during the Play mode (refer to Figure 23.22). The loading motor moves the idler gear up and attaches it to the top gear on the clutch, as shown in Figure 24.15. This will bypass the clutch; now you have direct drive to the spindles.

Removing Idler Wheels

In some models, you can remove the idler wheel and tire without removing the clutch. Also, you don't need to open up the undercarriage to remove a belt. There are several types of these idler wheel systems. Read the following sections and follow the instructions that pertain to the model you have.

FIGURE 24.15

The idler wheel attached to the top gear on the clutch.

First Type

1. In this type of system, remove the E-ring from the shaft on top of the idler wheel (see Figure 24.8).

2. Refer to the section in Chapter 7, "Removing E-rings."

3. Slide the idler wheel away from the drive wheel to clear the bracket above the drive wheel and simultaneously pull straight up on the idler wheel for removal. A small washer is under and on top of the idler wheel shaft. Don't lose it.

In other models:

1. Remove the spring connected on top of the idler wheel, as shown in Figure 24.16.

2. Remove the two mounting screws on each side of the bracket above the drive wheel and the reel motor, as shown in Figure 24.17. These same two screws mount the reel motor to the transport.

3. Let the reel motor drop down and slide the idler wheel assembly toward the front of the unit for removal, as shown in Figure 24.18.

The spring located on top of the idler wheel.

The mounting screws on each side of the bracket above the reel motor in the undercarriage.

FIGURE 24.18

The idler wheel and the idler wheel arm assembly.

When remounting:

1. Place the drive wheel arm assembly back into the wide part of the hole and slide the arm forward so that part of the arm is on top of the transport and the other part goes under the transport.

2. Remove the bottom plate (refer to Chapter 9, "Getting into the undercarriage").

3. Position the unit on the table so that you can grab the reel motor in the undercarriage.

4. Place the bracket back on top of the transport (see Figure 24.17) and align the mounting screw holes in the reel motor up with the holes in the bracket.

5. Remount the screws.

6. Remount the spring.

Second Type

This type of system has a small white clip at the base of the idler arm assembly, right above the drive wheel. Pull this clip straight back to release the latch, as shown in Figure 24.19. Pull the entire assembly straight up and off. To remount this assembly, reverse the process.

Releasing the latch to remove the idler arm assembly.

Third Type

This type of system is mounted with an O-ring on top of the shaft, right above the drive wheel on the idler arm assembly, (See Figure 24.20). Remove the O-ring. Refer to the section in Chapter 7, "Removing

Idler arm assembly with an O-ring.

O-rings." In a few models, a spring might be in the way, as shown in Figure 24.20. Remove this spring and slide the idler arm straight up and off the shaft. To remount this assembly, reverse the process.

Removing Idler Wheels and Clutch Assemblies

In these models, the idler wheel and the clutch are mounted to the same assembly. The idler wheel uses a rubber tire. If the cassette carriage is in the way, refer to the section in Chapter 8, "Removing and servicing cassette carriages." If you need to remove the transport, refer to the section in Chapter 9, "Transports." Several different methods are described in the following sections. Read each section and follow the proper instructions.

First Type

In this system, you do not need to remove the cassette carriage to remove the idler arm and clutch assembly.

1. Open the undercarriage.
2. Remove the belt from the pulley on the bottom of the clutch assembly (refer to Figure 10.2).
3. Flip the unit over and remove the Phillips head screws, one on each side of the idler arm and clutch assembly, as shown in Figure 24.21.
4. Lift the entire assembly straight up and out, as shown in Figure 24.22.

In a few models with the same type of idler arm and clutch assembly, the pulley that is mounted to the clutch in the undercarriage must be removed first. The reason is the pulley won't fit through the hole in the transport.

1. Take off the belt and remove the E- or O-ring on top of the pulley.
2. Refer to the sections in Chapter 7, "Removing E-rings and O-rings."
3. Pull the pulley straight off.
4. Turn the unit around and remove the mounting screws on each side to the idler arm and clutch assembly (same as in this section).
5. Lift the entire assembly straight up and out.

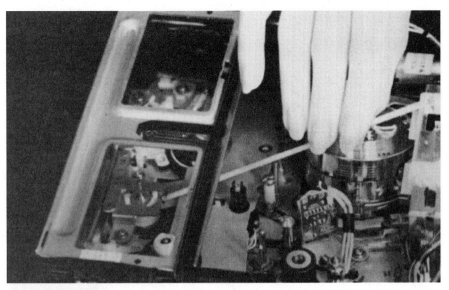

FIGURE 24.21

Locating the two Phillips head screws.

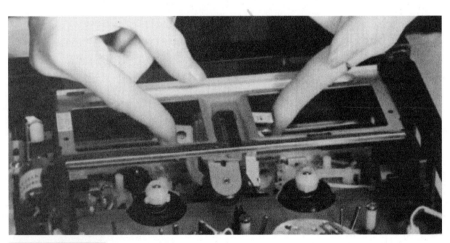

FIGURE 24.22

Removing the entire assembly.

FIGURE 24.23

The mounting screws to the idler wheel and clutch assembly.

If, after removing the idler wheel and clutch assembly, you need a new tire or clutch assembly, refer to the section in Chapter 1, "Parts."

Second Type

In this type of system, you need to remove the cassette carriage to remove the idler arm and clutch assembly.

1. Open the undercarriage.
2. Remove the belt from the pulley on the bottom of the clutch assembly (refer to Figure 10.2).
3. Flip the unit over and remove the five Phillips head-mounting screws, as shown in Figure 24.23 (I drew arrows pointing at the mounting screws). In a few models, the clutch assembly is very similar, but there are three mounting screws (instead of five) and the screws are mounted in a triangular pattern.
4. Remove the two white levers, one on each side of the clutch assembly, as shown in Figure 24.24 (I drew arrows pointing at the levers).
5. As you remove the levers, be sure to note how the pins on the bottom of the levers fit into the assembly.

The levers on each side of the idler wheel and clutch assembly.

6. To remove the lever on the left, slide the top of the lever over to clear the bottom of the brake shoe band adjustment and lift the lever straight up and off its shaft.

7. To remove the lever on the right, a small latch is above the lever where the arrow is pointing. Pull the latch back and lift the lever straight up and off its shaft.

8. On the other side of this lever is another lever that is attached. Leave it connected and slide both levers over to the side.

9. Lift the entire idler arm and clutch assembly straight up and out.

10. Flip the assembly over and slide the idler arm and wheel away from the clutch, as shown in Figure 24.25.

11. Now, you can replace the tire.

12. To remove the clutch, remove the O-ring above the pulley, then lift the pulley and the clutch straight up and off the shaft.

13. If you need a new tire or clutch, refer to the section in Chapter 1, "Parts."

FIGURE 24.25

Pointing at the tire on the idler wheel (left), clutch (right), and pulley (top).

To remount this assembly, reverse the process. However, a pin protrudes from the bottom of the assembly, as shown in Figure 24.26. This pin attaches to a sliding arm in the undercarriage. Slip the pin into the arm while you are remounting the assembly. When remounting the two levers, be sure to put the pin on each lever in the correct slot on the idler wheel and clutch assembly.

Third Type

In this type of system, the shaft of the idler wheel protrudes through the top of the transport between the two spindles (see Figure 24.7).

1. Remove the take-up spindle by removing its O-ring (a small washer is under the spindle—don't lose it).

2. Remove the small spring attached to the center of the shaft protruding through the transport (see Figure 24.7).

3. Open the undercarriage and remove the belt attached to the pulley at the bottom of the clutch assembly.

FIGURE 24.26

The pin that connects to the sliding arm in the undercarriage.

4. Remove the spring that is attached to the black cover and connected to a long white arm on the other end, as shown in Figure 24.27.

5. Wires run across the black cover through clips. Loosen the wires and pull them out of the clips.

6. Remove the black plastic cover, which is held on with one screw and two clips, as shown in Figure 24.28.

7. Remove the two O-ring's connected to the idler arm assembly and clutch assembly pulley, as shown in Figure 24.29. Some models have the same assembly, except that they have no black cover.

Follow the same procedures, omitting instructions about the black cover.

1. Slide the clutch assembly arm to the left where the larger hole is in the curved slot (see Figure 24.7).

2. Hold the idler wheel and pull the assembly and the shaft out through the hole in the slot.

FIGURE 24.27

The spring on the back cover.

FIGURE 24.28

Locating the screw and the two black clips.

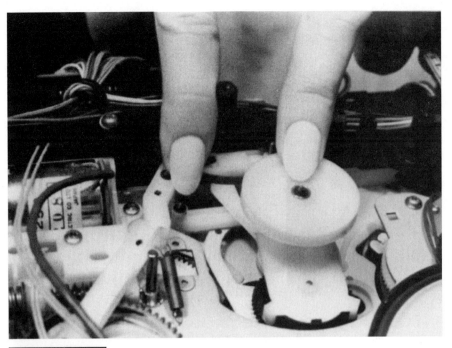

FIGURE 24.29

Locating the two O-rings for removal.

3. Some models do not have a larger hole in the slot on the transport. In these models, place pressure with a small flathead screwdriver, on the brass sleeve at the center of the pulley where the O-ring was (refer to Figure 24.29).

4. Pull the pulley straight back toward the screwdriver until the pulley pops off the shaft.

5. On the shaft underneath the pulley is a sandwich consisting of a washer, a gear, and another washer. Remove these three parts.

6. Pull the remaining clutch assembly off. The arm on the clutch bends, but (don't worry), it won't break.

7. Slide the idler wheel to the left at the end of the slot.

8. Pull off the idler arm and the idler wheel.

9. After removing the idler wheel and clutch assembly and you need a new tire or clutch assembly, refer to the section in Chapter 1, "Parts."

To remount this assembly, reverse the process. If you hear a grinding noise in the Fast-Forward or Rewind modes after reassembling the unit, one of the washers is incorrectly placed on the shaft of the clutch pulley.

Review

1. In the Fast-Forward or Rewind mode, if the unit runs slow or stops, hold the proper spindle and check for slippage between the drive wheel, idler wheel, and each spindle.

2. Clean each spindle and drive wheel. Clean the rubber tire on the idler wheel. Replace the tire, if necessary.

3. If fast forward or rewind doesn't work, look at the idler wheel to see if it's turning and check for positive contact against the spindles.

4. If the drive wheel doesn't turn, check the undercarriage for a broken belt, bad DC motor, or a bad clutch.

5. If the drive wheel isn't turning and the capstan shaft is turning, remove the transport and check for a bad belt or clutch.

6. To order a new belt, a DC motor, or a clutch assembly, refer to the section in Chapter 1, "Parts."

Capstan Shaft Problems

The capstan shaft, in conjunction with the pinch roller, is used to pull the video tape through the tape path. The capstan shaft is attached to the flywheel in the undercarriage. The flywheel is attached to the capstan motor. If the capstan shaft freezes or if the lubrication dries, it can cause the unit to run slow or not at all in the Play, Fast-Forward, and Rewind modes.

Detecting a Capstan Shaft Problem

1. Remove the main top cover (refer to Chapter 2, "Getting inside a VCR").

2. Insert a video cassette and push Play. The unit immediately loads up the video tape onto the tape path. The TV screen alternates between a blue display and a picture, or the picture appears to be in pause. After 30 to 60 seconds, the unit unloads and goes into the Stop mode.

3. Push Play again. Right after the unit loads the video tape, check to see if the reels inside the video cassette have stopped turning.

FIGURE 25.1

The video tape between the capstan shaft and the pinch roller.

4. If they have stopped, locate the capstan shaft and pinch roller. When the unit is in Play, the video tape goes between these two parts, as shown in Figure 25.1.

5. Push Play and watch the capstan shaft. Check to see if the shaft is turning. If the capstan shaft is turning but the pinch roller isn't turning, the pinch roller isn't pressed up tightly against the capstan shaft. If this tension on the video tape is missing, the capstan shaft is unable to pull the video tape through the tape path. The problem is with the pinch-roller assembly. The pinch-roller bracket could be bent, caught up on something, or the tension spring is missing. On the other hand, the pinch-roller gear assembly could be out of alignment. Refer to the sections in Chapter 27, "Removal of pinch-roller brackets" and "Straightening out pinch-roller brackets." Refer to the section in Chapter 15, "Pinch-roller alignment."

6. Another problem could be in the Play mode. The unit loads the video tape onto the tape path and unloads after 60 seconds, as previously explained. The difference is when the unit unloads the video tape it won't retract the tape back inside the video cassette. When you eject the video cassette, the video tape will hang up on the roller guides and crinkle the tape. The problem is that the capstan shaft isn't turning. There are several different reasons why this would happen. In older models, a belt around the outside of the capstan shaft flywheel might be broken or slipped off the flywheel or the capstan motor might be bad. Refer to the section in this chapter, "Checking a belt-driven VCR." Finally, the capstan shaft might have frozen. Refer to the sections in this chapter on "Removing the capstan shaft in belt-driven VCRs" or "Removing the capstan shaft in direct-drive VCRs."

Squeaking or Squealing in Play

In all models, the VCR is working okay, but has a constant loud squeak or squeal in the Play and Record modes. Sometimes this squeal starts immediately and sometimes it might start after 60 minutes. In some units, the same constant squeal will occur in Fast-Forward and Rewind modes. In other units, the squeal will stop in Fast-Forward and Rewind modes, or the squeal will pulsate. The problem is the bearings in the capstan shaft are dry and need to be lubricated. You'll need to remove the capstan shaft for lubricating.

Checking an Older, Direct-Drive VCR

1. Place the unit into the Stop mode. Most older direct-drive VCRs have an oil well that surrounds the base of the capstan shaft, as shown in Figure 25.2. If your unit has an oil well, you should be able to spin the capstan shaft. If the shaft turns freely, then the problem is the motor. Refer to the sections in Chapter 19, "Checking motors for dead spots" and "Checking for dead motors."

2. If the shaft won't turn or is difficult to turn, spray degreaser into the slot around the top of the oil well, as shown in Figure 25.3.

3. The degreaser frees the bearings by removing any rust or dirt. The first few turns of the shaft will be difficult, but keep turning the shaft until it turns freely.

4. After you've freed the shaft, lubricate the oil well with household oil until the felt in the slot is well saturated, but not overflowing.

5. Turn the shaft until it spins freely.

6. Be sure to wipe any excess oil off the capstan shaft.

7. Insert a video cassette and push Play.

8. Leave the unit in the Play modes for one or two hours so that the bearings lubricate properly.

FIGURE 25.2

An oil well surrounding the base of the capstan shaft.

FIGURE 25.3

Spraying degreaser into the oil well.

The fly-wheel.

Checking a Belt-Driven VCR

Older VCRs are belt driven. These VCRs don't have an oil reserve at the base of the capstan shaft. You can't turn the shaft by hand.

1. Unplug the VCR, place the unit on its side, and remove the bottom cover plate. If necessary, open the main circuit board to reveal the undercarriage. Refer to Chapter 9, "Getting into the undercarriage."

2. Look for the largest wheel in the undercarriage. This wheel is called the *flywheel*, as shown in Figure 25.4.

3. The flywheel has a belt wrapped around the outside of the flywheel and is attached to the pulley on the capstan shaft. Loosen the flat ribbon belt around the flywheel by pulling the belt off the pulley on the capstan motor, leaving the flywheel disconnected from the motor.

4. Place your finger on the outer edge of the flywheel and spin it (Figure 25.4). If it turns easily, the flywheel doesn't need to be lubricated. If the belt isn't broken, the capstan motor is bad. Refer to the sections in Chapter 19, "Belt-driven capstan motors", "Checking motors for dead spots" and "Checking for dead motors."

5. If the capstan motor is okay, the problem is in the circuits. If the flywheel is difficult or impossible to turn, you'll need to remove the capstan shaft to lubricate it. Refer to the sections in this chapter, "Removing the capstan shaft in belt-driven VCRs" and "Removing a frozen capstan shaft."

Checking Newer, Direct-Drive VCRs

Direct-drive systems are used in all newer models. To identify a direct-drive system, refer to the section in Chapter 19, "Direct-drive capstan motors." In some models, the capstan shaft is exposed (Figure 25.2) and you should be able to freely spin the shaft when the unit is in the Stop mode. In other models, an enclosure is wrapped half way around the back side of the capstan shaft (refer to Figure 4.6). This type of unit is inaccessible and you cannot spin the capstan shaft without first removing the transport. Refer to the section in Chapter 9, "Transports."

1. Locate the flywheel under the transport. A belt is on a pulley attached to the center of the flywheel (refer to Figure 10.2).
2. Remove the belt and place your finger on the outer edge of the flywheel and spin it. If it turns easily, the flywheel doesn't need to be lubricated and the system is malfunctioning; it has a circuit problem and you'll need to take it to your local service center for repairs.
3. If the flywheel is difficult or impossible to turn, you'll need to remove the capstan shaft for lubricating.
4. Refer to the section in this chapter, "Removing the capstan shaft in direct-drive VCRs."

Removing the Capstan Shaft in Belt-Driven VCRs

In older models:

1. Remove the screws at each end of the bracket, at the base of the flywheel, as shown in Figure 25.5.
2. Remove the bracket and the drive belt.

FIGURE 25.5

The fly-wheel bracket and its mounting screws.

3. Work the flywheel out by twisting it back and forth as you pull up on it. Continue until the flywheel is pulled out approximately ½ inch, then push the flywheel back in.

4. Look to the other side of the transport at the capstan shaft. A washer is approximately ½ inch up the capstan shaft, as shown in Figure 25.6.

5. Remove the washer from the capstan shaft. Be sure to put the washer in a safe place. The washer keeps the oil from coming up the capstan shaft and getting on the video tape.

6. Return to the undercarriage and pull the flywheel and the capstan shaft all of the way out by twisting and pulling. If you fail to follow this procedure, the washer on the capstan shaft will fall off into the transport and might become lost.

FIGURE 25.6

A washer halfway up the capstan shaft.

Removing the Capstan Shaft in Direct-Drive VCRs

In newer models, you need to remove the transport.

1. Refer to the section in Chapter 9, "Transports."

2. Remove the belt (refer to Figure 10.2).

3. To remove the capstan shaft and flywheel, read the following descriptions and choose the one that fits your unit.

A. In some models, remove the E-ring at the base of the capstan shaft, as shown in Figure 25.7 (refer to the section in Chapter 7, "Removing E-rings").

B. In other models, remove the E-ring at the top of the capstan shaft, as shown in Figure 25.8.

C. In a few models, release the clip at the base of the capstan shaft by placing an ice pick behind the clip that goes into the groove and push up on the bottom of the clip with the aid of a small screwdriver, as shown in Figure 25.9.

FIGURE 25.7

E-ring at the base of the capstan shaft.

FIGURE 25.8

E-ring at the top of the capstan shaft above the enclosure.

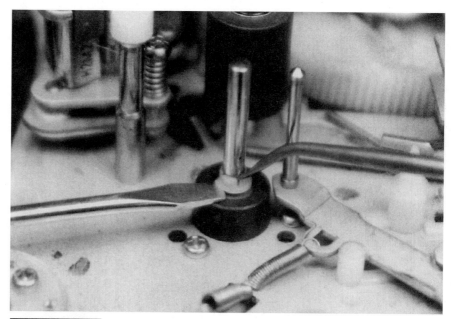

FIGURE 25.9

Removing the clip at the base of the capstan shaft.

    D. In newer models, remove the metal harness that protrudes over the flywheel and is mounted to the transport with one Phillips head mounting screw, as shown in Figure 25.10.

    E. In a few models, no E-ring, clip, or bracket holds the flywheel down. Just remove the tightly fitted white washer at the base of the capstan shaft (refer to Figure 25.6).

To remove the capstan shaft:

1. Work the flywheel out by twisting it back and forth as you pull on it. When you first pull on the flywheel, it will be pulling back because of the magnets inside of the flywheel.

2. Continue until the capstan shaft is pulled out approximately ½ inch, then push the flywheel back in.

FIGURE 25.10

Mounting harness over the fly-wheel.

3. Look to the other side of the transport at the capstan shaft.
   A. In some models, a washer is approximately ½ inch up the capstan shaft (Figure 25.6). Remove the washer from the capstan shaft.
   B. Other models have no washer to remove. Newer models have a cover wrapped half way around the back side of the capstan shaft (refer to Figure 4.6). There will be two white washers, one at the base and the other at the top of the shaft, right under the enclosure, as shown in Figure 25.11 (I moved the washers out so that you can see them). Twist and pull on the flywheel until the capstan shaft almost comes out all the way, then push the shaft back in approximately ⅔ of the way. Turn the transport around and slide the two washers up and off the shaft.

4. Twist the flywheel back and forth as you pull the capstan shaft out of the capstan shaft hole. Now, you can see the complete capstan motor with its coils and the magnets under the flywheel, as shown in Figure 25.12.

FIGURE 25.11

Washers at the base and top of the capstan shaft.

FIGURE 25.12

The coils and magnets under the fly-wheel.

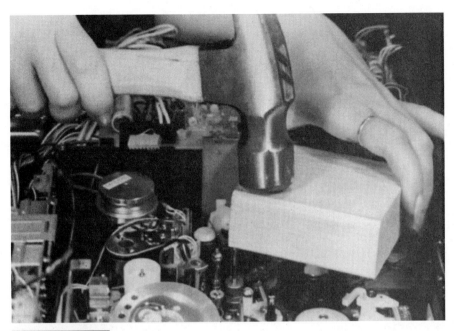

FIGURE 25.13

Tapping down the capstan shaft with a hammer and a block of wood.

Removing a Frozen Capstan Shaft

In older models only:

1. If the flywheel and the capstan shaft are frozen tight, remove the flywheel bracket and drive belt (refer to Figure 25.5).

2. Try to remove the flywheel assembly by twisting the flywheel back and forth, pulling on it at the same time.

3. If it doesn't move, place the unit right side up.

4. Place a block of wood on top of the capstan shaft, as shown in Figure 25.13.

5. Tap the block lightly with a hammer to loosen the capstan shaft for removal.

6. Place the unit back on its side and remove the flywheel and capstan shaft, as previously explained.

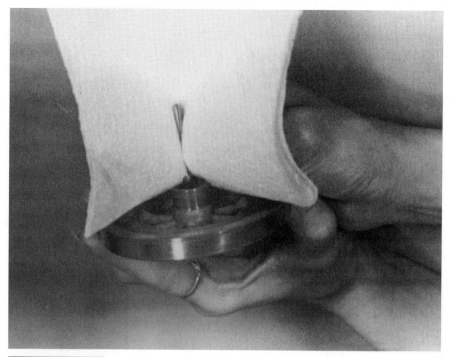

FIGURE 25.14

Cleaning the capstan shaft.

Cleaning and Lubricating the Capstan Shaft

In all models:

1. Dip the corner of a paper towel into some cleaning solution.

2. Place the capstan shaft onto the wet portion of the towel, as shown in Figure 25.14.

3. Twist it back and forth while pulling the capstan shaft out of the paper towel. This procedure removes all the dirt and residue from the capstan shaft. You might need to repeat this step several times.

4. With a can of degreaser, using the extended nozzle, spray degreaser into the capstan shaft hole, as shown in Figure 25.15.

5. Place the glass brush into the hole. Refer to the section in Chapter 1, "Making a glass brush."

6. Twist the brush back and forth as you push it all the way through.

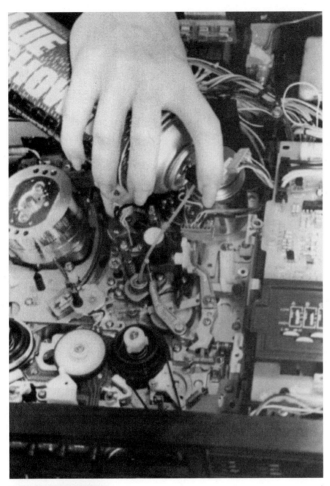

FIGURE 25.15

Spraying degreaser down into the capstan shaft hole.

7. Pull the brush out, spray it with degreaser, and clean the brush off with a paper towel. Repeat this procedure until all the residue inside the hole is removed.

8. After you've completed this process, spray out the hole to remove any fibers left by the glass brush.

In models where the capstan shaft is exposed (refer to Figure 4.5):

1. With a can of household oil, place the lubricant on the capstan shaft only.

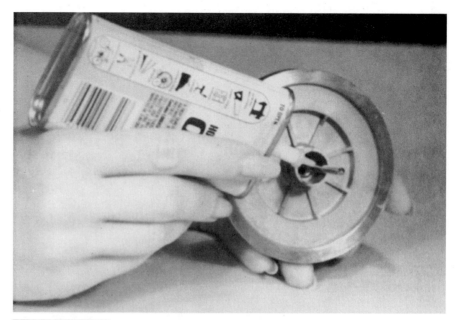

FIGURE 25.16

Lubricating only the lower portion of the shaft.

2. Lubricating from the flywheel where the capstan shaft is attached, up the shaft about an inch, as shown in Figure 25.16. Don't get any oil on the upper portion of the capstan shaft where the video tape makes contact.

3. Place the capstan shaft back into the capstan shaft hole.

4. Push the flywheel in all the way.

5. Leave the unit on its side and spin the flywheel. It should spin freely by itself for a minute or so. If it still turns a little hard or unevenly, repeat the procedure.

In newer models that have two washers on the capstan shaft (see Figure 25.11), start the lubricating point at the flywheel on the capstan shaft and going up the shaft about ¾ inch.

1. Place the capstan shaft back into the capstan shaft hole.

2. Push the flywheel in until the shaft is protruding out the other side approximately 1 inch, then stop.

3. Turn the transport over and place both washers onto the capstan shaft.

4. Push the washers down the shaft approximately ½ inch.

5. Place a drop of oil onto the top end of the capstan shaft.

6. Push the flywheel in all the way.

7. Push one washer to the base of the shaft and the other to the top on the shaft under the enclosure with the aid of a screwdriver. Figure 25.11 shows the washers, but they are not in the correct position.

8. Leave the transport on its side and spin the flywheel. It should spin freely by itself for 10 seconds or longer.

Remounting, Greasing, and Adjusting the Capstan Shaft

In older models:

1. Before replacing the mounting bracket, look on the inside of the bracket (see Figure 25.5). A plastic plateau is in the center of the bracket. It usually is round in shape.

2. This plateau has old grease or no grease at all. Using a paper towel, remove the old grease and apply a small amount of phono lube onto its center.

3. Look at the center of the flywheel. A shaft protrudes out about ¼ inch. This is the base of the flywheel. With a paper towel, wipe off all the old grease on the end of the shaft.

4. Replace it with new grease, as shown in Figure 25.17.

5. After cleaning, lubricating, and greasing the flywheel and capstan shaft, it's time to remount the drive belt onto the outside of the flywheel.

6. Place the mounting bracket back into the base of the flywheel and replace the two mounting screws.

7. Some models have an adjustment screw on the bottom of the bracket, under the plateau, as shown in Figure 25.18. Don't touch this adjustment screw, unless it is necessary. You can determine

whether an adjustment is needed by moving the flywheel up and down. It should have no more than ¾ to 1 mm of play. If it's too tight, the capstan shaft won't turn freely. If it's too loose, the capstan shaft moves up and down eating the video tape.

8. Remount the bottom cover plate and circuit board, if needed.

9. Place the unit right side up.

10. Replace the washer you set aside earlier. Slide it down the capstan shaft to the base of the shaft.

11. Use a dry paper towel to wipe off the capstan shaft. It's important to wipe off any excess oil. Any excess oil will get on the video tape and clog the video heads.

Newer models have no bracket at the bottom of the flywheel to be greased. Remount the E-ring at the base of the capstan shaft (refer to

FIGURE 25.17

Placing grease on the base of the fly-wheel.

FIGURE 25.18

The fly-wheel adjustment screw.

Figure 25.7). The clip at the base of the capstan shaft (refer to Figure 25.9) or remount the harness holding the flywheel down (refer to Figure 25.10). Place the washer back onto the capstan shaft at its base.

Review

1. To find out if you have a capstan shaft problem, insert a video cassette and push Play. Check to see if the video tape is moving. If not, check the capstan shaft to see if it's turning.

2. If you have an older direct-drive VCR and it turns hard or is frozen, use degreaser and lubricate the oil well at the base of the capstan shaft.

3. If you have a belt-driven VCR, remove the belt and spin the flywheel to check its freedom of movement. If the flywheel turns hard or is frozen, remove the bottom bracket, the capstan shaft, and flywheel for cleaning and lubricating.

4. If you have a newer direct-drive VCR, spin the capstan shaft to see if it spins freely. If the capstan shaft is inaccessible, remove the transport and spin the flywheel.

5. Be sure to remove the washer or washers around the capstan shaft before completely removing the shaft.

6. Clean the capstan shaft with a paper towel and cleaning solution. Clean the capstan shaft hole with a glass brush.

7. In a belt-driven VCR, be sure to replace the grease on the plateau of the mounting bracket and the bottom of the flywheel.

8. Before reassembling the capstan shaft, be sure that you place oil on the shaft, but don't get any oil on the upper portion of the capstan shaft where the video tape makes contact.

9. Be sure to place the washer or washers back onto the capstan shaft.

Jammed Cassette Carriages

The cassette carriage in your VCR can become jammed for several reasons. You will not be able to insert or retrieve your video cassette. The reasons why and how to remove a jammed video cassette are covered in this section.

The Reasons Why a Cassette Carriage Jams

There are eight different reasons why a cassette carriage can jam. Read each of the reasons and use the one that fits your situation.

First Reason

You might push Eject and the video cassette starts to come up, but suddenly it stops and goes back down, or the video cassette might almost come completely out, but stop and go back in. On the other hand, you might try to insert a video cassette and it comes right back out. Two reasons could cause this problem. It could be a bad mode switch (refer to the section in Chapter 21, "Mode switches") or you might find a foreign object caught in the teeth of the gears or worm gear that makes the cassette carriage move.

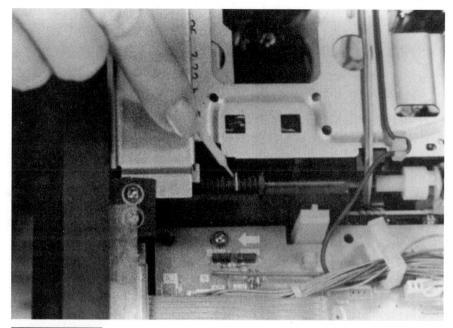

FIGURE 26.1

A foreign object caught in the gearing.

Look closely at these gears. A very small piece of plastic, wood, or metal can jam between two teeth or in the worm gear, as shown in Figure 26.1. The reason the gears jam is because the gears usually have grease on them and when a very small foreign object falls onto the gear, it'll stick to it. When the gears move across the foreign object, the object keeps the gears from turning. The objects can be as small as a pin head.

In older models:

1. Insert a blank cartridge into the cassette holder (refer to Figure 1.1). Note: If you don't have a blank cartridge, then use a video cassette.

2. Look to the back on the right side of the cassette carriage for a drive pulley where the drive belt is connected.

3. Place your finger on the drive pulley and rotate it, as shown in Figure 26.2. The pulley should turn in a clockwise direction.

4. As you rotate the pulley, push gently on the blank cartridge until it is all the way inside of the carriage, then the pulley will take over. All the gears will turn and the cassette holder will move.

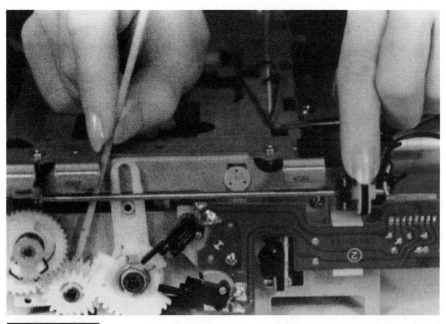

FIGURE 26.2

Turning the pulley to make the gears rotate.

5. When the gears no longer turn, look where the teeth come together on each gear on both sides of the cassette carriage. Make a mental note of this position. This is where you'll find the foreign object.

6. If you cannot see the gear clearly at this time, remove the carriage.

7. Refer to Chapter 8, "Removing and servicing cassette carriages."

8. After removing the carriage, turn the pulley counterclockwise to back up the gears exposing the foreign object for removal.

In some models a worm gear is attached directly to the motor located on the right of the cassette carriage, as shown in Figure 26.3. Use your finger or the tip of a flathead screwdriver to rotate the base of the worm gear. Follow the same procedure as described in the paragraph above for checking the gears.

In newer models, all cassette carriages are directly geared to the transport. You will need to remove the carriage before checking the gearing. You will find the drive gear at the bottom right side of the carriage, as shown in Figure 26.4, or the last gear on the right side of the carriage, as shown in Figure 26.5. Use your finger to rotate the drive gear and follow the preceding procedure to check the gears.

FIGURE 26.3

The worm gear attached to a housing loading motor.

Some newer models have a plastic or a metal shaft that runs along the bottom of the right side of the cassette carriage that slides back and forth. The front part of this shaft has a row of teeth, as shown in Figure 26.6, which connect to the other gears on the cassette carriage. Push the end of the shaft. When the shaft moves, the gears will move. Follow the preceding procedure to check the gears.

Second Reason

You try to insert a video cassette into your VCR or you push Eject and try to remove a cassette already inside the VCR. Instead, you hear the

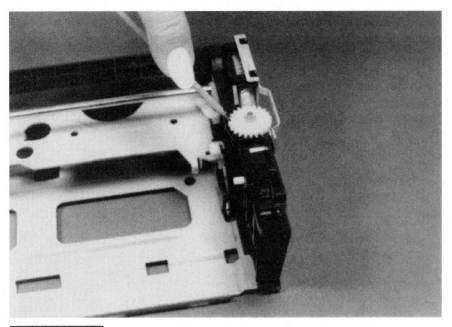

A drive gear on a cassette carriage.

Another type of drive gear on a cassette carriage.

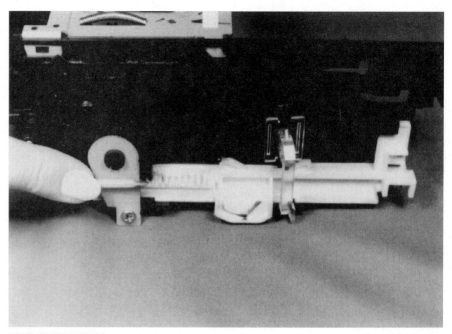

FIGURE 26.6

Sliding gear or arm on a cassette carriage.

sound of a motor running inside the unit, but no video cassette comes out or goes into the unit and then the sound stops. In a few models, the unit will shut down.

The problem is a broken loading motor belt. In older models, this belt is located on a pulley (Figure 26.2) near the rear of the cassette carriage.

In newer models, the belt and pulley are located on the loading motor at the top right side, at the back of the transport, as shown in Figure 26.7. In either case, just replace the broken belt. Proceed to the first four sections of Chapter 10 and to the section in Chapter 19, "Removing loading motors and belts," for detailed instructions.

Third Reason

You insert a video cassette and hear a squealing sound. The tape comes back out.

This condition is caused by a drive belt that has been stretched out and is slipping on the pulleys. Replace the belt. To locate these belts, refer to the previous section.

FIGURE 26.7

Loading motor drive belt at the rear of the transport.

Fourth Reason

You insert a video cassette and receive absolutely no response.

This condition can be caused by a bad cassette housing motor, loading motor, or a bad tape sensor. Refer to the sections in Chapter 19, "Checking motors for dead spots" and "Checking for dead motors." Also, refer to the section in Chapter 21, "Tape sensors."

Fifth Reason

You insert a video cassette, but it only goes in part way and then it jams. All the mode buttons are nonfunctional, so you unplug the power and then plug it back in to reset the microcomputer. The video cassette will automatically eject. On the other hand, if the video cassette doesn't automatically eject, unplug the power. Then you'll have to manually turn the pulley or the worm gear to eject the video cassette.

Refer to the section in this chapter, "Manually removing video cassettes." After retrieving the video cassette and while the unit is still unplugged, insert the video cassette by placing a finger on each side of the back of the cassette and pushing it in. The cassette should insert

FIGURE 26.8

Cassette holder shaft going through the arm of a gear.

evenly. If one side goes in easily and the other locks, a gear is probably broken on one side of the carriage. Refer to Chapter 8, "Removing and servicing cassette carriages," and the section in this chapter "Locking levers and latches." After removing the carriage, look for:

1. A bad gear with missing or worn off teeth.

2. A plastic shaft connected to the cassette holder protruding through a slot on the arm of the loading gear and the shaft of the cassette holder is missing (broken off) or part of the plastic arm attached to the gear itself is broken off (see Figure 26.8).

3. A tab that holds the tension spring against the cassette holder shaft is broken off (see Figure 26.9).

4. A gear has a crack running across it and the shaft it's attached to will not turn with the gear because of the crack in the gear.

5. Two gears, side by side, are hooked together with springs between the two gears. A tab holding a spring is broken off (see Figure 26.10). One gear will turn and the other will not. To remove and replace the broken gear, refer to the section in Chapter 15, "Cassette carriage alignment."

FIGURE 26.9

A plastic tab attached to a spring going to the arm of the gear.

FIGURE 26.10

A plastic tab attached to a spring between two gears.

Sixth Reason

You insert a video cassette and it goes in smoothly. Later, you push Eject and the video cassette jams up against the inside of the cassette carriage door. After five seconds, the video cassette reloads.

The reason is the cassette carriage door doesn't open. I have found certain instances in which, for no reason, the eject lever which opens the carriage door, has slipped or has been pushed behind the tab on the door. To correct this problem, you will need to remove the front cover and remount it (refer to the first three sections in Chapter 32).

Seventh Reason

You insert a video cassette and the cassette jams up and comes right back out or it might stay in the unit. If it stays in the unit and you push Play, the unit will jam up, then shut down. You unplug the power and then plug it back in to reset the microcomputer. The video cassette automatically ejects.

The problem is that the door-lock release for the video cassette is bent or that the coil spring behind it has fallen out. Figure 26.11

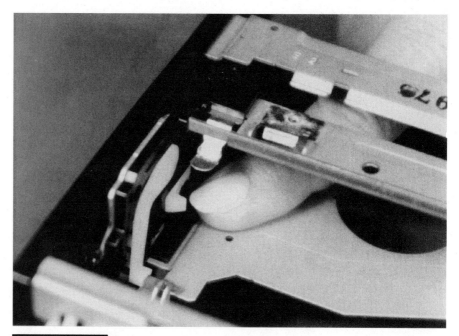

FIGURE 26.11

A pin that pushes on the release button of the video cassette.

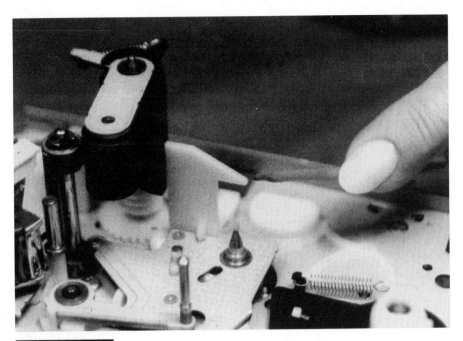

Door opener for the video cassette.

shows the pin that pushes on the release button to the door of the video cassette (refer to Figure 29.1). The other part that can cause this problem is the door-opening bracket, as shown in Figure 26.12. If it is made of metal, it can become bent or misadjusted. If it is made of plastic, it can become misadjusted or broken and the video cassette can jam up against it, or miss it entirely. Another type of door opener assembly has a spring behind the door opener, as shown in Figure 26.13. Sometimes the spring will pop off its tab and causes the door opener to be in the wrong position when you insert a video tape.

In newer models, the door opener is the protruding slanted piece in the front of the loading motor housing bracket or the protruding slanted piece on the lower pinch-roller bracket assembly (refer to Figure 15.13). In these units, the problem is that the housing bracket or lower pinch-roller assembly hasn't been remounted properly.

Eighth Reason

You've inserted a video cassette in crooked and the cassette holder started to pull the tape in. Because it was inserted crooked, the video

FIGURE 26.13

Spring behind the door opener.

cassette didn't move with the cassette holder. Eject won't eject the tape.

The cause of this problem is that the locking latches have locked in the cassette carriage. To correct this problem, unplug the power and plug it back in so that the microcomputer resets. The tape then automatically ejects.

Locking Latches and Levers

The cassette holder will not drop down without a video cassette inserted.

The cassette holder is locked down on both sides by latches. These latches have two locking positions; they are the two square holes on each side of the cassette carriage. These latches are spring loaded and will pop up into the holes which keeps the cassette holder from moving, as shown in Figure 26.14. Two locking levers release the latches located at the rear of the cassette holder. One is shown in Figure 26.15.

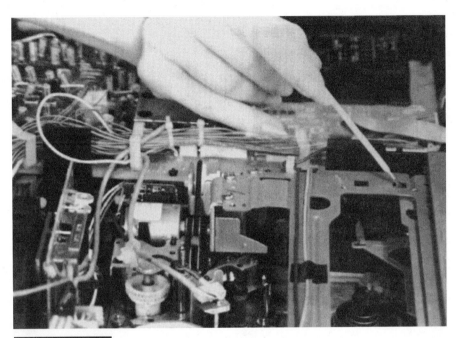

FIGURE 26.14

The locking latch protruding through the hole in the carriage.

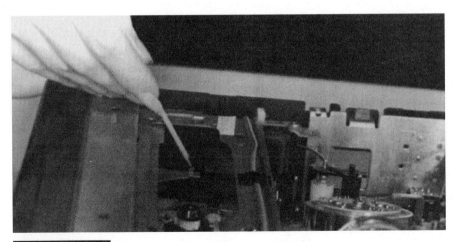

FIGURE 26.15

The locking lever.

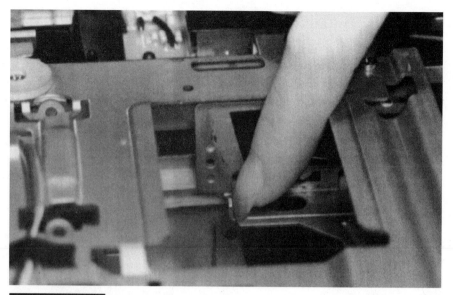

FIGURE 26.16

The locking lever on the back edge of the cassette holder.

The other locking lever is parallel on the opposite side of the cassette holder.

In other models, the two levers are on the very back edge of the cassette holder, as shown in Figure 26.16. A video cassette pushes against these levers when inserted. When these locking levers are pushed forward, the locking latches drop down out of each hole on the carriage, unlocking the cassette holder. Sometimes the locking lever gets bent back and the locking latch doesn't unlatch, not allowing the cassette holder to move forward when inserting a video cassette. Straighten out the locking latch so that it has a 90-degree angle.

Manually Removing Video Cassettes

A video cassette is in the unit and eject won't eject it.

Remove the main top cover. The video tape is inside the video cassette and not loaded or wrapped around the video drum.

In older models, locate the pulley at the right rear corner of the cassette carriage. This pulley (Figure 26.2) has a drive belt attached. Turn the pulley counterclockwise and continue this process until the video cassette ejects. You might find a worm gear, instead of a pulley, located on the right side of the carriage (Figure 26.3). Rotate the base of the worm gear counterclockwise until the video cassette ejects.

In newer models, the cassette carriage is directly connected to the transport by gears and the loading motor retracts the roller guides and ejects the video cassette. In 70% of all newer models, the loading motor is located at the right rear corner on top of the transport (Figure 26.7). Rotate the pulley or worm gear counterclockwise until the video cassette ejects.

In other new models, the loading motor is located under the transport and is not accessible. In this case, you need to pull the cassette carriage out. The carriage can be removed with the video cassette inside of it (refer to Chapter 8, "Removing and servicing cassette carriages"). Then, rotate the drive gear or push on the sliding gear on bottom of the cassette carriage to remove the video cassette (see Figures 26.4 and 26.6).

Removing a Video Cassette Stuck in Play Mode

In older models, to remove a video cassette stuck in the Play mode, you need to retract the roller guides. To do this:

1. Locate the loading motor. A loading motor is the only motor attached to the transport and is horizontal to it and can be located on top of the transport or in the undercarriage. Refer to the section in Chapter 19, "Loading Motors."

2. Check the top of the transport for the loading motor (refer to Figures 19.2 and 19.3). If not, refer to Chapter 9, "Getting into the undercarriage."

3. Place the unit on its side and locate the loading motor (refer to Figures 10.5, 19.21, and 19.22).

4. Rotate the pulley or worm gear on the loading motor counterclockwise until the roller guides on top of the transport have completely retracted.

5. Retract the loose video tape back into the video cassette.

6. Locate the flywheel in the undercarriage connected to the capstan shaft (see Figure 10.2).

7. Turn it in either direction until it pulls the video tape back into the cassette. Then, follow the procedure in this chapter, "Manually removing video cassettes."

FIGURE 26.17

Loading motor and worm gear under transport.

In newer models:

1. Remove the transport. Refer to the section in Chapter 9, "Transports."

2. Locate the loading motor on top of the transport (refer to Figure 19.5) or the loading motor under the transport, as shown in Figure 26.17.

3. Rotate the pulley or worm gear counterclockwise until the roller guides on top of the transport have completely retracted.

4. Retract the loose video tape back into the video cassette.

5. Locate the flywheel under the transport (refer to Figure 10.2) and turn it in either direction.

6. Go back to the loading motor and rotate it again until the video cassette ejects.

Review

1. Always insert a video cassette in straight, not crooked.

2. If you can't insert a video cassette or the video cassette doesn't eject, check for a broken or bad drive belt, then look for foreign objects caught in the teeth of the gears.

3. You try to insert a video cassette, the cassette seems to keep going in crooked and jamming up, then ejects. Check the cassette carriage for a broken gear.

4. You insert a video cassette and it comes right back out. Check the automatic door release, the door opening bracket, the locking latches and levers, or the mode switch.

5. When manually removing video cassettes and the unit is stuck in Play, first turn the pulley on the loading motor, then turn the flywheel. Then, in older units, turn the pulley on the cassette carriage. In newer units, continue turning the pulley on the loading motor.

Video-Tape-"Eating" VCRs

It's so frustrating to rent a movie, pop some popcorn, and insert the movie for viewing, only to have the VCR suddenly "eat" the video tape. Sound familiar? This chapter covers where and why this annoying problem happens and how to remove the damaged tape from your VCR.

Two Locations at Which a Video Tape Can Be Eaten

Your VCR is most likely to eat a video tape at two different locations. The first location is between the pinch roller and the capstan shaft. The other location is on the take-up tape guide. This section shows you how to remove the tape if it is caught in either of these two locations.

First Location

The video tape goes between the pinch roller and the capstan shaft, as seen in Figure 27.1 (I've taken off the top cover and door of the video

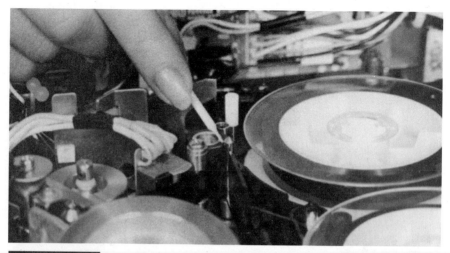

FIGURE 27.1

Video tape between the pinch roller and capstan shaft.

cassette to give you a better view—you don't need to remove these parts). If the pinch roller and the capstan shaft aren't aligned exactly parallel to each other, the video tape will be eaten.

Second Location

Another location that a video tape can be eaten is on the take-up tape guide. This tape guide is between the A/C head and the capstan shaft. Normally, the video tape flows smoothly across the tape guide. In some cases, however, pressure is on the video tape, pushing it against the bottom edge of the tape guide. This pressure causes a crease or a fold in the video tape. The fold travels between the capstan shaft and the pinch roller, causing the bottom edge of the video tape to be eaten. The bottom edge of the take-up guide causes the fold, as shown in Figure 27.2. The A/C head is not properly aligned with the video tape. The misalignment puts pressure on the tape in a downward motion. Do not touch the adjustment nut on top of the take-up tape guide. Push Play and locate the Allen nut or adjustment screw located directly behind the A/C head, as shown in Figure 27.3. This tilt-adjustment screw causes the A/C head to change its contact angle with the video tape by either tilting the A/C head forward or backward.

The tilt adjustment is either a Phillips-head screw or a 1.5-mm Allen nut. Place the proper tool in the rear nut or screw and begin turning it counterclockwise. As you slowly turn the tilt adjustment,

Second location at which the video tape can be eaten.

the A/C head starts tilting backward and changing its contact angle with the video tape. Keep an eye on the bottom of the tape guide where the wrinkle is. At some point, the video tape will move up the tape guide and smooth out.

If you have the TV monitor on at this time, watch the picture being produced from the video tape. A horizontal belt of lines will appear across the screen. This belt of lines start at the bottom of the screen and moves up toward the top of the screen, and then starts over again. You can correct these two problems by turning the tilt-adjustment screw clockwise slowly until the bottom edge of the video tape is just touching the bottom protruding edge on the tape guide, as shown in Figure 27.4. Now proceed to the section in Chapter 13, "Tilt adjustment."

> **QUICK»TIP**
>
> If you turn the tilt adjustment too fast, the video tape could pop up across the tape guide and position itself too high. If you overadjust the tilt, the video tape could start wrinkling at the top of the tape guide.

FIGURE 27.3

Locating the tilt adjustment on the A/C head.

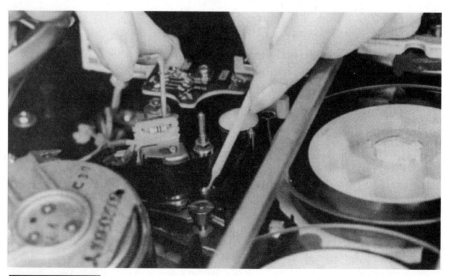

FIGURE 27.4

Aligning the video tape to the take-up tape guide.

Do not touch any other adjustments on the A/C head. If you do, you could lose the audio and video synchronization. If you accidentally turn the wrong screw and one of these problems exists, proceed to Chapter 13, "Audio head alignment."

Reasons Why a VCR Will Eat a Video Tape

There are several reasons why your VCR will suddenly decide to eat one of your video tapes. These reasons range from residue buildup to a bad pinch roller. This section shows you how to solve these problems quickly and easily.

First Reason

I've taken a capstan shaft assembly out of a unit to show you what to look for. Look at the upper and lower portions of the capstan shaft, as shown in Figure 27.5. You're looking for residue buildup. Residue

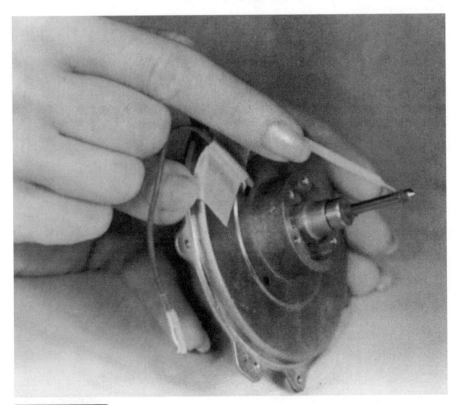

FIGURE 27.5

Locating residue build-up on the capstan shaft.

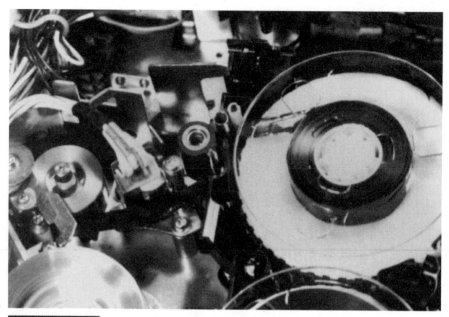

FIGURE 27.6

Video tape being eaten between the pinch roller and capstan shaft.

buildup keeps the pinch roller from sitting flush against the capstan shaft. The pressure on the pinch roller is incorrect and causes the video tape to be pulled up or down the shaft and to be eaten.

Look at the video tape as it passes between the capstan shaft and pinch roller while the unit is playing. The tape is being crinkled up and eaten, as shown in Figure 27.6.

To avoid this problem:

1. Clean off the residue buildup on the capstan shaft.

2. Immerse a chamois stick or a glass brush into cleaning solution.

3. Once it's fully saturated, clean the capstan shaft, as shown in Figure 27.7.

4. Apply pressure with the cleaning utensil; the residue buildup can be quite hard.

5. To clean the pinch roller, refer to the section in Chapter 4, "Pinch roller" under the subhead "Cleaning the various components."

FIGURE 27.7

Cleaning the capstan shaft.

Second Reason

The bracket that holds the pinch roller can also cause the video tape to be eaten. This bracket can be bent or twisted, causing uneven pressure between the pinch roller and the capstan shaft. The point where it might get bent or twisted is shown in Figure 27.8.

1. Check for residue build-up on the capstan shaft or pinch roller.

2. If it is clean, insert a video tape and push Play.

3. Observe the video tape as it goes between the capstan shaft and the pinch roller. See if the video tape is being pulled down and is crinkling at the bottom edge of the tape. If so, take a small screwdriver and place it at the top edge of the pinch-roller cap, as shown in Figure 27.9.

4. Apply a small amount of pressure, pushing away from the capstan shaft. Do not push too hard or the pinch roller will stop turning. This should cause the video tape to straighten out between the

pinch roller and the capstan shaft, but only when the pressure is being applied. On the other hand, if the video tape is being pulled up and is crinkling at the top of the video tape, put pressure toward the capstan shaft, as shown in Figure 27.10.

5. Some models don't have a pinch roller cap. To apply the pressure, place the corner of a flathead screwdriver blade onto the center of the plastic pin inside of the pinch roller, as shown in Figure 27.11.

6. If the pinch roller is located upside down, place your screwdriver in the center of the metal bracket, right where the rivet is, above the pinch roller, as shown in Figure 27.12. Use this method to move the video tape up or down the pinch roller. This procedure should straighten out the video tape.

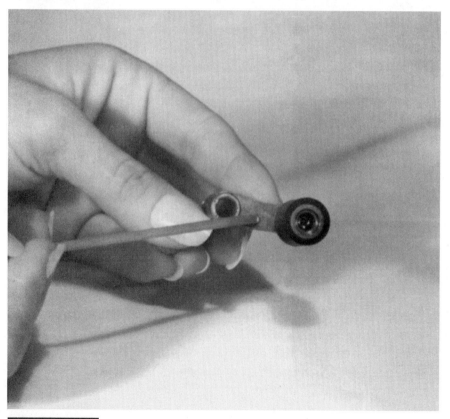

FIGURE 27.8

The point the pinch roller might get bent or twisted.

FIGURE 27.9

Pushing the top center of the pinch roller toward the back of the VCR.

FIGURE 27.10

Pushing the top center of the pinch roller toward the front of the VCR.

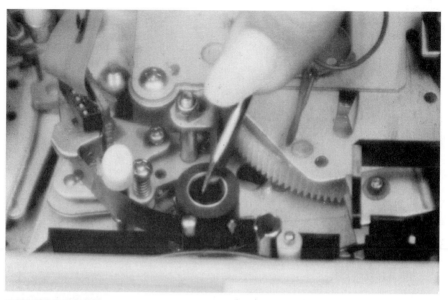

FIGURE 27.11

Pushing on the pin inside the pinch roller.

FIGURE 27.12

Pushing on the rivet above the pinch roller.

In either case, if the video tape straightens out while pressure is being applied, then the pinch-roller bracket is bent or twisted.

Third Reason

A pinch roller can go bad two different ways. The pinch roller might become hard and shiny, like a piece of hard plastic, instead of hard rubber. The video tape will become unstable, move up and down on the capstan shaft, and be eaten. The pinch roller also can become warped and cause uneven pressure on the video tape. The fluctuating pressure causes the tape to rise or fall and to be eaten. In either case, replace the pinch roller.

To check for this type of problem, remove the pinch roller and place a screwdriver shaft almost up against the pinch roller. Look for any gaps or bows in the pinch roller, as shown in Figure 27.13. If the pinch roller is inaccessible, you will need to remove the pinch roller. Refer to the section in Chapter 7, "Removing a pinch roller." If the pinch roller is bowed, for replacement, refer to the section in Chapter 1, "Parts."

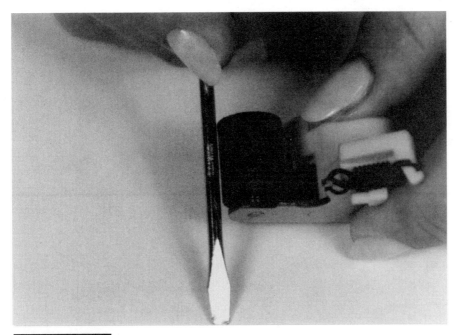

FIGURE 27.13

Showing the gap between a straight edge and the pinch roller.

FIGURE 27.14

A ball of video tape wrapped around the capstan shaft.

Fourth Reason

You're watching a rented movie and suddenly the VCR speeds up, then the unit stops playing. You try to push the Play button again, but the unit still won't work. You push the Eject button, but the video cassette doesn't eject.

Take off the main top cover. You'll probably find a ball of video tape wrapped around the capstan shaft, as shown in Figure 27.14. The reason this happens is that rental tapes take a lot of abuse from previous renters. The tapes can become sticky from spilled beverages or children's sticky fingers, causing the tape to stick to the capstan shaft and roll up in a ball.

A ball of video tape is usually wrapped around the capstan shaft and wedged tightly between the capstan shaft and the front cover of the video cassette, as shown in Figure 27.15 (I have removed the top cover of the video cassette to give you a better view).

To remove the video cassette:

1. Remove the ball of video tape wrapped around the capstan shaft.

2. Place a flathead screwdriver directly under the ball of video tape.

FIGURE 27.15

A ball of video tape wedged against the video cassette.

3. Twist the tip of the screwdriver back and forth between the base of the capstan shaft and the bottom of the ball of video tape. This works the ball of video tape up the capstan shaft.

4. Continue to work the ball of video tape up the shaft for at least ¼ inch.

FIGURE 27.16

Prying the ball of video tape up the capstan shaft.

5. Carefully find a solid bracket to lay the shaft of the screwdriver on and pry the ball of video tape up the capstan shaft as far as it will go, as shown in Figure 27.16.

6. The cassette holder will hold the video cassette down and keep the ball from coming all the way off the capstan shaft.

7. Push the Eject button and pry the ball of video tape up the shaft at the same time. Front-loading units are motor driven and the motor must be running to pry the video tape off.

You might need to repeat this procedure several times before the cassette holder will eject all the way. Be careful to place the shaft of the screwdriver on a solid bracket when prying on the ball of video tape. Do not place the screwdriver shaft on any part that can be bent or broken.

Usually, the ball of video tape comes off the capstan shaft fairly easily. Sometimes, however, the ball of video tape is wrapped around

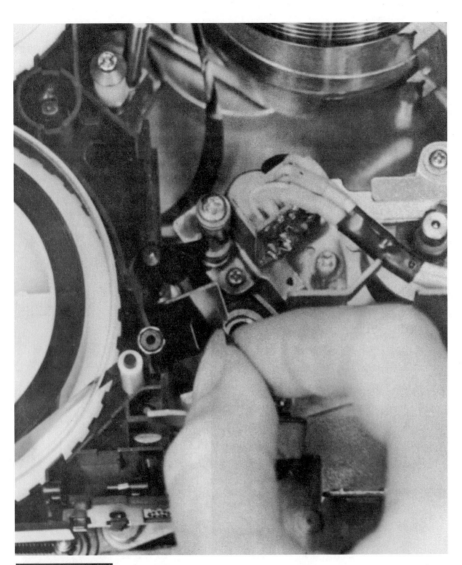

FIGURE 27.17

Slicing the ball of video tape with a razor blade.

the capstan shaft so tightly that you can't work it up the shaft with a screwdriver. In this case:

1. Cut the ball of video tape off the capstan shaft with a single-edged razor blade.

2. Slice it in the same place until you cut down to the metal shaft, as shown in Figure 27.17.

3. Push Eject and the video cassette ejects. Sometimes the cassette carriage won't eject because the remaining tape between the capstan shaft and the front of the video cassette jams it.

4. Simply place a screwdriver under the video cassette near the capstan shaft and pry up, as shown in Figure 27.18.

5. Be sure to push Eject at the same time you are prying up on the cassette. The video cassette should eject fairly easy with the aid of a screwdriver.

6. After removing the video cassette, finish unwrapping the rest of the video tape left on the capstan shaft.

7. Remove all of the cut pieces of video tape from the unit.

8. Refer back to Chapter 4, "Capstan shaft" under the subheading "Cleaning the various components" and to Chapter 29, "Repairing a video tape."

FIGURE 27.18

Prying up on the bottom of the video cassette.

Fifth Reason

A video cassette is in the unit and you push Eject. When the cassette comes out, it leaves about a foot or so of video tape stuck on a roller guide inside the unit. The tape might even come out with the cassette, but it's crinkled.

The problem is the supply spindle didn't pull the video tape back into the cassette when the VCR left the Play or Record mode. To correct this problem, refer to Chapter 23, "Take-up spindle problems." The faulty part is probably the idler wheel or clutch assembly. This part drives the supply spindle.

Removal of Pinch-Roller Brackets

In most models, the pinch-roller bracket consists of a very thin metal and bends easily, although the bracket varies in size and shape from model to model. This section includes the five major types of brackets for removing pinch-roller brackets.

After you remove the pinch-roller bracket, you have an option. You can order a new pinch-roller bracket (refer to the section in Chapter 1, "Parts") or you can straighten out the one you already have. Take the new or straightened-out bracket and place it back into its original position. To remount all brackets simply reverse the following procedures.

First Type

In this model, remove the O-ring and the tension spring, as shown in Figure 27.19, and then pull the bracket straight up and off. To remove O-rings, refer back to the section in Chapter 7, "Removing O-rings."

Second Type

In some models, additional parts might have to be removed, such as a mounting bracket or a small circuit board. These parts are held on with a screw.

1. Simply take out the screw, as shown in Figure 27.20, and remove the mounting bracket or the small circuit board, as shown in Figure 27.21.
2. Remove the tension spring. An easy way to remove the spring is by taking a stiff piece of wire or a paper clip and make a hook on the end of it.

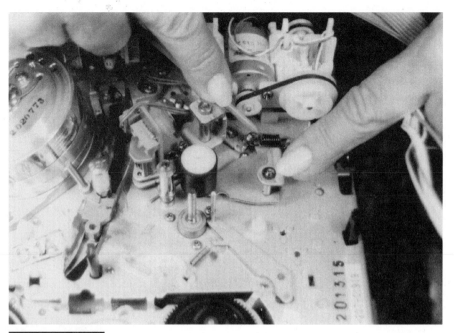

FIGURE 27.19

O-ring and tension spring.

FIGURE 27.20

A circuit board obstructing the pinch roller mount.

Removing the circuit board or bracket.

3. Place the hook through the hook on the end of the spring and pull the spring off the spring mount, as shown in Figure 27.22, or use long-nose pliers.

4. Place the spring out of the way.

5. The bracket has another arm attached to it.

6. You'll need to detach one end of the arm to remove the pinch-roller bracket. All attached arms are held on by a C-, E-, or O-ring.

7. Remove the ring and the attached arm comes up and off, as shown in Figure 27.23. Refer back to the section in Chapter 7, "Removing C-, E-, or O-rings."

8. After removing all the attached parts, remove the C-, E-, or O-ring that is holding down the pinch-roller bracket.

9. Pull the pinch-roller bracket straight up and off, as shown in Figure 27.24.

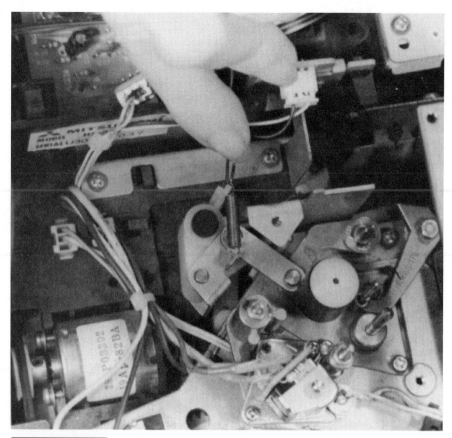

FIGURE 27.22

Pulling the spring off its mount.

Third Type

In this model:

1. Remove the cassette carriage (refer to Chapter 8, "Removing and servicing cassette carriages").
2. Remove the two O-rings on top of the housing over the pinch roller (refer back to Figure 7-19) and lift the housing straight up and off.
3. Remove the O-ring above the pinch-roller bracket, as shown in Figure 27.25, and remove the tension spring.
4. This bracket has a tracking arm attached to the cam gear and the arm slides along grooves on top of the cam gear.
5. Pull the pinch-roller bracket straight up and off. The tracking arm comes right off the cam gear. Be sure that the tracking arm is back in the groove when replacing it.

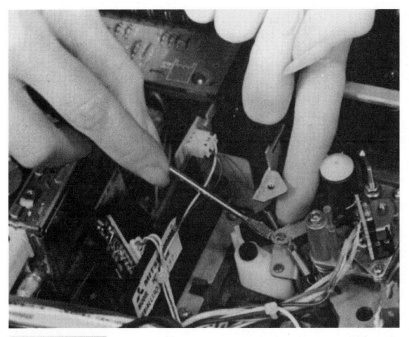

FIGURE 27.23

Removing the attached arm from its bracket.

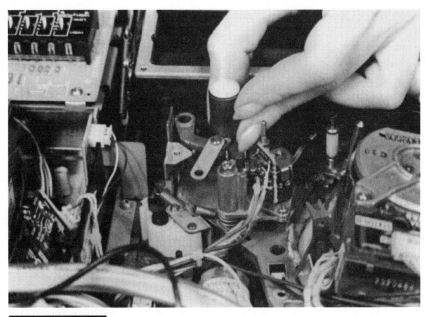

FIGURE 27.24

Removing the pinch roller and bracket.

The O-ring, tension spring, and tracking arm.

In another model, which has a tracking arm on the pinch-roller bracket:

1. Remove the cassette carriage (refer to Chapter 8, "Removing and servicing cassette carriages").

2. Remove the three screws on top of the housing above the cam gear, as shown in Figure 27.26, and remove the belt off the loading motor (I drew arrows pointing at the mounting screws).

3. Unplug the loading motor and lift the housing and motor straight up and off.

4. Remove the O-ring above the pinch-roller bracket, as shown in Figure 27.27, and pull the bracket straight up and off. In this case, the tension spring is incorporated within the bracket itself and comes off with the bracket.

5. The tracking arm will come right off the cam gear. Be sure that the tracking arm is back in the right groove when replacing it.

Loading motor, housing, and the three mounting screws.

O-ring and leaf spring attached to the pinch roller bracket.

Fourth Type

In newer models:

1. Remove the top portion of the pinch-roller assembly to remove the pinch-roller bracket. This type of pinch roller is mounted upside down. Note the position of the pinch roller before removing the assembly. Look for a latch at the top of the assembly (refer back to Figure 7-14) or the latch could be along the side of the pinch-roller assembly (refer back to Figure 7-15).

2. In either case, pull the latch straight back to clear the assembly and, at the same time, grab hold of the top of the assembly and pull it straight up and off.

Fifth Type

In some models, the pinch-roller bracket is made of plastic. If the bracket is warped, you must replace it.

Every model is different but this will give you a good idea of how to remove pinchroller brackets.

Straightening Out Pinch-Roller Brackets

1. Find the location on the pinch roller that will come in contact with the capstan shaft.

2. Place the back of the pinch-roller bracket flat against the T of the T-square.

3. Place the ruler against the pinch roller, where it would come in contact with the capstan shaft, as shown in Figure 27.28.

TRADE SECRET In models that have the assembly attached to the bracket, you need to remove the pinch-roller bracket from off the assembly before placing the bracket on a flat surface. In all other models, the brackets are already separated.

4. If the pinch roller is not perfectly flush against the ruler, then place the flat side of the bracket on a flat surface to determine where the bracket is bent.

5. Bend or twist the pinch-roller bracket until the bracket sits perfectly flat on the table. Recheck by using the T-square.

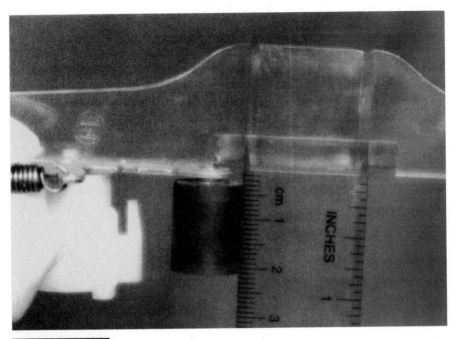

FIGURE 27.28

Using a T-square on the pinch roller bracket.

Review

1. Examine the following when looking for problems with the pinch roller and capstan shaft.

   A. Residue build-up on capstan shaft.
   B. A bent or twisted pinch-roller bracket.
   C. A bow in the pinch roller.
   D. The rubber on the pinch roller.
   E. A ball of video tape around the capstan shaft.

2. If a ball of video tape is on the capstan shaft, it's not the fault of the VCR, but is caused by a sticky substance that was left on the video tape.

3. When the tape is being wrinkled at the bottom of the take-up tape guide, do not touch the adjustment nut on top of the tape guide. Replace the take-up clutch assembly or turn the A/C head tilt adjustment.

4. When adjusting the A/C head, be sure that the video tape is flat against the tape guide and that the tilt adjustment is turned as far clockwise as possible without wrinkling the tape.

5. You're viewing a video and suddenly the VCR stops and you cannot eject the video cassette. Check the capstan shaft for a ball of video tape wrapped around the shaft.

6. When removing the video tape from the capstan shaft and you use the aid of a screwdriver, be careful not to bend or break any other parts.

Repairing Supply and Take-Up Spindle Systems

A VCR contains two spindles. One is located on the left side of the transport under the cassette carriage and is called the *supply spindle*. The other one is located on the right side of the transport under the cassette carriage and is called the *take-up spindle* (refer to Figure 19.13).

Functions of the Spindles

When you insert a video cassette, the cassette holder inside of the cassette carriage places the video cassette so that both reels inside the cassette will sit on top of the spindles (refer to Figure 23.2). The protruding shaft on top of each spindle inserts up inside each reel. The total weight of each reel, plus the video tape, rests entirely on top of the spindles. This gives complete control of the speed and the direction of the video tape, with no drag from inside of the video cassette.

Spindle Problems

Four types of problems with spindles can contribute to the cause of the video tape running slow, uneven, stopping, and making strange sounds. Read the following and choose the one that fits your problem.

FIRST PROBLEM

The first type of problem deals with older models. The spindles are driven by an idler wheel, the idler wheel uses a rubber tire, and the tire drives the spindles in Play, Fast-Forward, or Rewind (refer to Figures 24.3 and 24.4) modes. The problem is that the spindle can slip on the tire, causing the video tape to slow down or stop. To correct this problem, refer to the sections in Chapter 23, "Cleaning wheels and spindles made of plastic" and "Cleaning rubber tires."

SECOND PROBLEM

The second type of problem is that you hear a grinding or clicking sound when the VCR is in the Play, Fast-Forward, or Rewind modes. The reels inside of the video cassette might be jerking, slowing down, stopping, and restarting. The problem could be the teeth on the spindle, the idler gear, or drive gear are worn down or a couple of teeth are missing (refer to Figures 24.5 and 23.7). The problem also could be that the idler arm is bouncing on and off the take-up spindle. Check for a bad clutch. Refer to the section in Chapter 23, "Third type" under the subhead "Tape-up spindle systems." If the teeth on the spindle are bad, refer to the section in this chapter, "Removing spindles." If the teeth on the idler gear are bad, refer to the section in Chapter 24, "Removing idler wheels." For replacement parts, refer to Chapter 1, "Parts."

THIRD PROBLEM

The third type of problem deals with spindles mounted to reel motors. When you place your unit into the Play, Fast-Forward, or Rewind modes, you hear a rubbing sound, like plastic rubbing against plastic. One of the reels inside the video cassette is rubbing on the top or bottom of the inside of the cassette. The problem is that the spindle under that reel is set too high or too low. If the height of the spindle is too far off, it can keep the reel from turning. Reset the height on the spindle. Refer to the section in this chapter, "Removing spindles attached to a motor."

FOURTH PROBLEM

The fourth type of problem is you are in Fast Forward, or Rewind, and you push Stop. One of the reels inside of the video cassette keeps spinning and unrolling video tape. The problem is the brake shoe isn't putting enough pressure against the spindle. The tension spring on one of the brake shoe arms will be stretched out, missing, or discon-

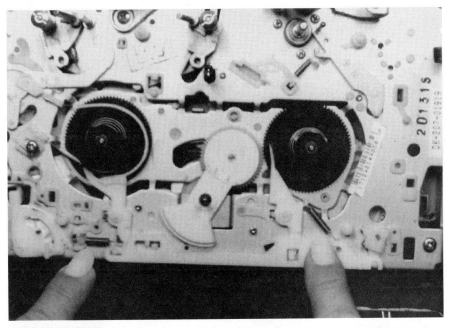

FIGURE 28.1

A coil spring attached to each brake shoe arm.

nected. Some units have one brake shoe on each spindle and other units have two brake shoes on a spindle. The springs could be a coil spring at the end of each arm, as shown in Figure 28.1, or it could be a leaf spring on the back of each arm, as shown in Figure 28.2. On the other hand, a coil spring could connect two brake shoe arms together with each arm going to the opposite spindle, plus two more brake shoe arms with a separate spring on each one, as shown in Figure 28.3 (I drew arrows pointing at the coil springs). You need to replace or reconnect the spring to the brake shoe arm.

Removing Spindles

Four mounting systems are used for spindles. Remove the cassette carriage (refer to Chapter 8, "Removing and servicing cassette carriages"). Read the following and choose the one that fits your unit.

FIRST TYPE

In some models, a C- or O-ring is at the top of the spindle. Remove the ring (refer to the sections in Chapter 7 "Removing C-rings" or

FIGURE 28.2

A leaf spring attached to each brake shoe arm.

FIGURE 28.3

Three coil springs attached to four brake shoe arms.

FIGURE 28.4

Lifting the brake shoe band over the supply spindle.

"Removing O-rings"). If the unit uses a C-ring, a small washer will be under the C-ring. Remove and place both pieces aside.

To remove the supply spindle:

1. Carefully pull the brake shoe band up and over the spindle and hold it there, as shown in Figure 28.4.

2. With the other hand, lift the spindle straight up and off its shaft.

3. Lower the band back over the spindle shaft, where the spindle was, to its original position. One to three small washers will be under the spindle. Sometimes one of these washers will stick to the bottom of the spindle. Be sure to check the bottom of the spindle.

4. When remounting the spindle, reverse the process, except after the spindle is back on its shaft and the brake shoe band is back around the spindle. You will notice that the brake shoe is spring loaded.

5. Place this brake shoe over the brake shoe band (see Figure 28.5).

6. Remount the small washer, C-, or O-ring.

Latch on top of the brake shoe arm.

7. To remove a take-up spindle, use the same procedure as those mentioned previously, but there isn't a brake shoe band.

8. When remounting, pull the brake shoe or brake shoes away from the spindle (refer to Figure 19.14) so that the spindle can slip all the way down its shaft.

9. Release the brake shoe or brake shoes up against the base of the spindle (see Figure 28.3).

SECOND TYPE

In other models, an O-ring is on top of the spindle. Also, a brake shoe arm is at the base of the spindle that is holding it down.

1. Remove the O-ring.

2. Remove the tension coil spring attached to the arm and release the latch located on top of the arm, as shown in Figure 28.5.

3. Lift the arm straight up and off.

4. In some models, you can move the brake shoe arm over to clear the spindle (refer to Figure 19.14).

5. Remove the O-ring and slide the spindle off its shaft. Use the same procedure as described in the first type of spindle removal.

THIRD TYPE

Newer models use the upper mounting bracket that holds the top portion of the brake-shoe assembly together (see Figure 28.7) or, in other models, the upper bracket that holds the top portion of the idler-wheel assembly together (see Figure 28.6). In either case, these brackets have two tabs, one over each spindle. The tabs are located at the base of each spindle, as shown in Figure 28.6.

1. Remove the two mounting screws on each side of this bracket.

2. Remove the three mounting screws on the bracket above the brake-shoe assembly, as shown in Figure 28.7 (I drew arrows pointing at the mounting screws).

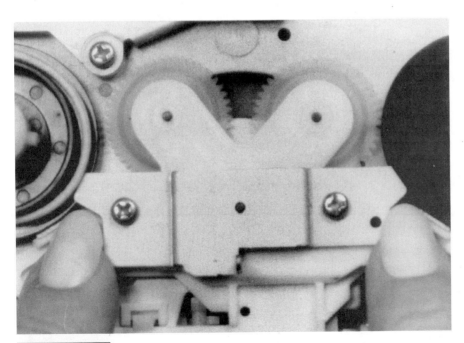

FIGURE 28.6

The tabs protruding over the spindle on the upper bracket of the idler wheel assembly.

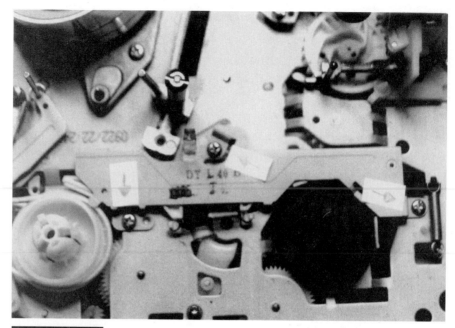

FIGURE 28.7

The mounting screws on the bracket covering the brake shoe assembly.

3. Lift either bracket straight up and off. In most cases, the brake shoe assembly or idler-wheel assembly will stay on the transport. Try not to disturb these parts. If you flip the transport over, some of these parts can fall out. In some models, you will need to remove the brake shoe arm as well.

4. Remove the tension spring fastened to the arm.

5. Look for a latch on top of the arm (see Figure 28.5) or look for a small square hole in the transport where part of the arm protrudes through, as shown in Figure 28.8.

6. This part of the arm has a clip fastened to it. With a small flathead screwdriver, push in on this portion of the arm to release the clip in the undercarriage and simultaneously lift the arm up and out.

7. Follow the same procedure as described in the first type of spindle removal.

FIGURE 28.8

The clip that protrudes through the hole in the transport.

FOURTH TYPE

In other models, the brake shoe arms hold the spindle down.

1. Remove the two mounting screws on each side of the idler-wheel assembly (see Figure 28.9).

2. Slide the idler wheel assembly over so that the tab on top of the assembly will clear the sliding bracket over the brake-shoe arm assembly, as shown in Figure 28.9.

3. Two double latches are near each end of the sliding bracket, as shown in Figure 28.10. Squeeze one latch together. Pick up that end of the sliding bracket just enough to clear the latch.

4. Release the other latch and lift the bracket straight up and off.

5. Remove the springs attached to each brake-shoe arm (refer to Figure 28.3).

6. Lift each arm up and off the shafts. Be sure to remember which arm went on which shaft.

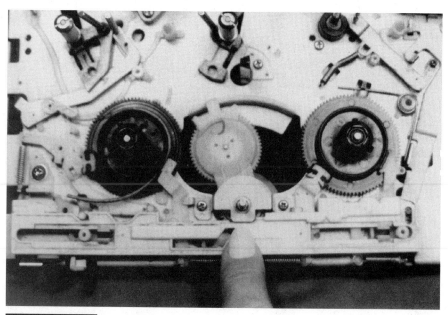

FIGURE 28.9

Pointing at the tab on the idler wheel assembly above the sliding bracket.

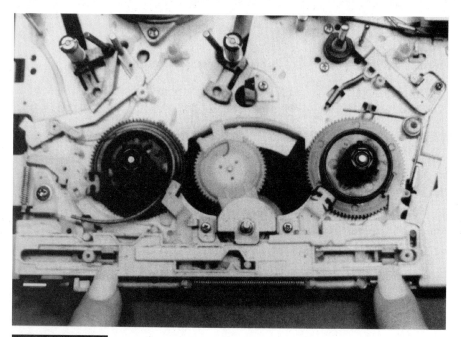

FIGURE 28.10

Double latches on the sliding bracket above the brake shoe arm assembly.

7. Another clip might be connected to one of the arms (see Figure 28.8).

8. Follow the same procedure as described in the first type of spindle removal.

Removing Spindles Attached to a Motor

Two mounting systems are used for spindles. One system has two lock nuts at the base of the spindle. The other system has an O-ring at the top of the spindle. Look for an O-ring on the top of the spindle to determine which type of system you have.

FIRST TYPE

The first system has an O-ring located on top of the spindle.

1. Remove the O-ring, then pull the spindle straight up and off.

2. After removing the spindle, you'll see a brass base connected to the motor shaft on which the spindle was sitting.

3. Measure the height from the top of the DC motor to the top of the brass base, as shown in Figure 28.11. Write this measurement down; you'll need it to realign the spindle.

4. Remove the brass base by loosening both Allen lock nuts located on each side.

5. Pull the brass base straight up and off the shaft.

6. When remounting the brass base onto the motor shaft, reposition the base to its correct measured height, as previously shown.

7. Tighten down the lock nuts.

8. Look at the bottom of the spindle. It's the same shape as the outside of the brass base. Line up these two patterns when remounting the spindle. If they are not aligned properly, the spindle won't slide down all the way.

9. Remember to push aside the brake shoe (refer to Figure 19.14). The brake shoe can keep the spindle from sliding down all the way.

10. Place the O-ring over the top of the shaft. With the aid of a small flathead screwdriver, push the O-ring down securely in the groove near the top of the shaft. If not properly aligned, the O-ring won't fit or reach the groove.

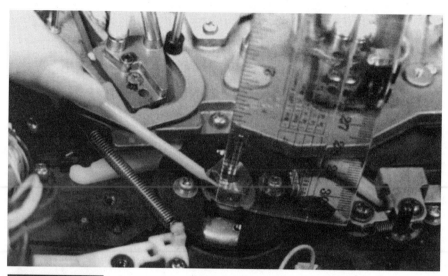

FIGURE 28.11

Measuring from the top of the motor to the top of the brass base.

SECOND TYPE

For spindle systems using base lock nuts:

1. Measure the height from the transport to the top edge of the base of the spindle, as shown in Figure 28.12. This measurement is very important.

2. Write down the measured distance, for reference, when remounting it. Two lock nuts hold the spindle onto the motor shaft. These nuts are located directly under the spindle.

3. With a 1.5-mm Allen wrench, loosen both lock nuts by turning them counterclockwise, as shown in Figure 28.13.

4. Pull the spindle straight up and off.

5. After removing the spindle, you will be able to see what the two Allen set screws look like and where they're located on the base of the spindle, as shown in Figure 28.14.

To remount the spindle on the motor shaft:

1. Insert an Allen wrench into one of the lock nuts and leave it there.

2. Slide the spindle down the shaft.

FIGURE 28.12

Measuring the distance from the transport to the spindle.

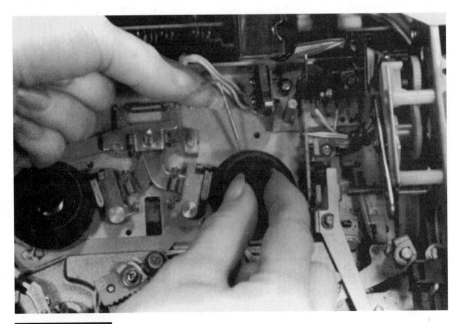

FIGURE 28.13

Loosening the lock nuts.

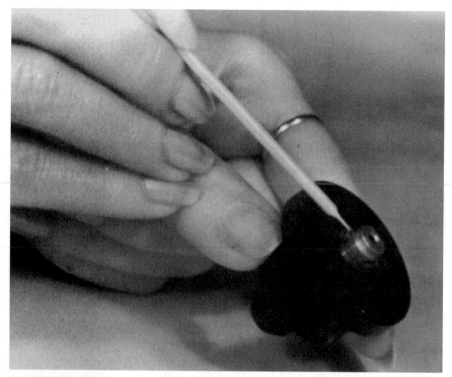

FIGURE 28.14

Locating the lock nuts on the spindle.

3. As you slide the spindle down the shaft, locate the brake shoe and pull it away from the spindle. The spindle must pass by the brake shoe.

4. Tighten the lock nut that holds the Allen wrench until it's snug. Leave it loose enough to move the spindle up and down the shaft.

5. Remeasure the distance between the transport and the top of the base of the spindle, as previously shown in Figure 28.12.

6. Position the spindle to the correct measurement and tighten down the lock nut.

7. Remove the Allen wrench, insert it into the other lock nut, and tighten it. It's important that the measured distance is exact. If the measurement is off, it causes the reel inside of the video cassette to rub on the inside of its own cover.

To remove and remount the take-up spindle, follow the same procedure as previously explained. The exception is, in older models, that the take-up spindle might have a tape counter belt attached to it.

1. Keep the belt tight when pulling it off the spindle.

2. Slip the belt over any bracket it will reach so that the other end of the belt doesn't come off the pulley on the tape counter. In most models, the tape counter pulley is hard to reach.

3. If the belt comes off the pulley, you might have to remove the front cover to remount it.

4. When replacing the belt onto the take-up spindle, be sure the belt is in the groove.

Review

1. The total weight of each reel and video tape rest on top of the spindles, making the reels free from any drag.

2. Spindles can contribute to the cause of the video tape running slow, uneven, stopping, and making strange sounds.

3. Push Stop. One of the reels inside of the video cassette keeps spinning and unrolling video tape. Check the tension spring connected to the brake-shoe arm.

4. The spindles are held down by a C-, O-ring, brake-shoe arm, or by tabs on the upper bracket over the idler wheel or brake-shoe arm assemblies.

5. When removing the supply spindle, carefully pull the brake-shoe band up and over the spindle.

6. After removing a spindle, always check the bottom of the spindle for a washer stuck to it.

7. When removing a spindle attached to a motor, be sure to measure the height before removing the spindle.

8. When replacing a spindle attached to a motor, be sure to remeasure it for proper height.

Repairing a Video Tape

According to Murphy's law, it never fails, somehow the special video tape of a child's birthday party or Uncle George's 75th anniversary has been eaten by the VCR.

What usually happens to a video tape:

1. The capstan shaft crinkles the video tape.

2. The video tape sticks to the capstan shaft or pinch roller, wadding it into a ball.

3. You break the video tape trying to remove it from the VCR.

4. One of the broken ends gets rewound back into the video cassette where you can't reach it.

> **QUICK ›› TIP**
>
> To remedy all of these dilemmas, carefully follow the following instructions. It's important to remember that grease, dirt, or any oil on your hands, can rub off onto the video tape. Wash and dry your hands thoroughly before touching the video tape.

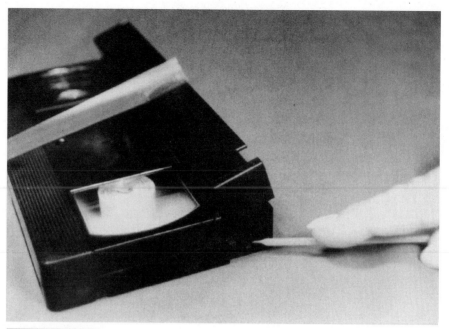

FIGURE 29.1

Taping back the door and the release button on the video cassette.

Opening a Video Cassette

1. Face the front of the video cassette toward you.

2. You'll find a small button on the left side, just behind the door.

3. Push this button to release the door on the front of the cassette.

4. Pull the door open all the way.

5. Tape the door back with a piece of scotch tape, as shown in Figure 29.1.

6. Flip the cassette over and remove the five screws on the bottom of the video cassette, as shown in Figure 29.2 (I drew arrows pointing at the five screws).

7. Check the sides and back of the video cassette for any labels. If the labels have been on the cassette for a long period of time, they might be difficult to remove.

8. With a single-edge razor blade, slice each label down the middle, following the crack along the sides and back of each video cassette, as shown in Figure 29.3.

FIGURE 29.2

The five screws on the bottom of the video cassette.

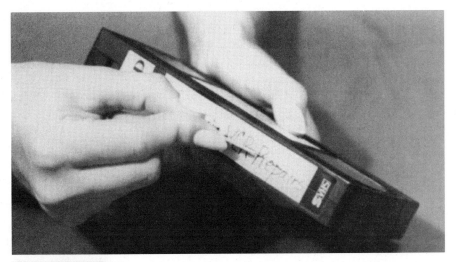

FIGURE 29.3

Slicing the label with a razor blade.

FIGURE 29.4

Lifting off the top cover.

9. Hold both the top and bottom covers at the same time.

10. Turn the cassette right side up and lift off the top cover, as shown in Figure 29.4.

Unlocking the Reels

After removing the top cover, you can see two exposed reels inside the cassette, as shown in Figure 29.5. When looking at each reel, notice that teeth are on the outside of each reel. These teeth are locked down by lock arms to keep the reels from turning when the cassette is inactive. To release the reels, pull the lock arms away from the teeth; this allows the reels to turn freely, as shown in Figure 29.5.

Threading the Video Tape

1. Face the video cassette toward you. This side of the video tape broke off and pulled up inside the cassette, as shown in Figure 29.5.

2. Pull the take-up reel, which is the reel on the left, straight up and out.

3. Pull a foot or so of video tape out of the reel, as shown in Figure 29.6.

FIGURE 29.5

The reels and locking arms inside the cassette.

FIGURE 29.6

Pulling the video tape out of the reel.

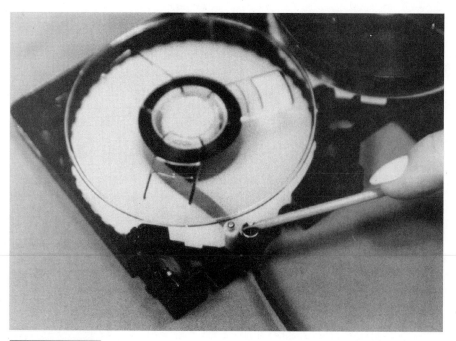

FIGURE 29.7

Placing the video tape between the take-up guides.

4. Lower the reel close to the cassette and place the pulled out video tape between the white plastic roller guide and the metal guide, as shown in Figure 29.7.

5. Lower the reel back down into its original position.

6. Look at the supply reel, the reel on the right-side of the cassette. On this side, the video tape has been eaten and crinkled.

7. Release the lock arm and pull out some good video tape.

8. Look at the roller guides on this side. The video tape goes between a metal pin and a fiber spring, then around a metal guide. In some models, the metal guides have a white plastic roller around them. In Figure 29.8, the fiber spring is being pulled away, to give you a better view.

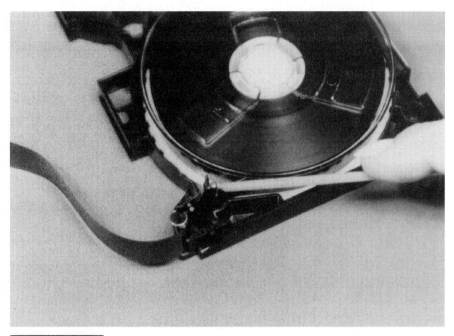

FIGURE 29.8

Placing the video tape between the supply guides.

Splicing a Video Tape

1. Cut off all the crinkled up video tape, as shown in Figure 29.9, leaving only smooth video tape coming out from each reel.

2. Place the two smooth ends of video tape, overlapping them by about three inches.

3. Be sure that each end of the video tape is perfectly aligned. In other words, both pieces should be laying on top of each other, appearing as one piece.

4. Be sure that the video tape coming from each reel isn't twisted. With a pair of sharp scissors, cut both pieces of video tape at a slight angle, as shown in Figure 29.10. By cutting the video tape in this manner, the ends will match up perfectly.

5. Get a roll of ¾ inch wide clear tape and tear off a piece approximately two inches in length.

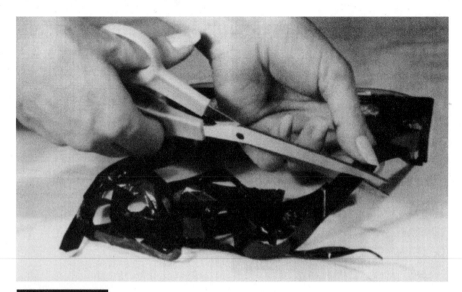

FIGURE 29.9

Cutting off the crinkled video tape.

FIGURE 29.10

Aligning and cutting the video tape.

6. Hold the piece of clear tape in one hand with the sticky side up.

7. Take one of the ends of the video tape and place the side that will not come in contact with the video heads, approximately two-thirds of the way across the clear tape.

8. Attach the video tape to the clear tape at this point, as shown in Figure 29.11.

9. Lay down the two attached pieces of tape with the sticky side up.

10. Take the other end of video tape coming out of the video cassette and place the backside of it on top of the taped piece laying on the table. Be careful. Do not let the top piece of video tape touch the clear tape yet.

11. Reposition the video tape so that both sides are perfectly aligned, as shown in Figure 29.12.

12. Gently slide the top piece of video tape toward the cut end of the bottom piece until the top piece drops off the end.

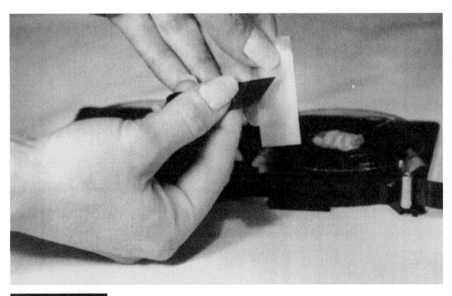

FIGURE 29.11

Attaching the video tape to the scotch tape.

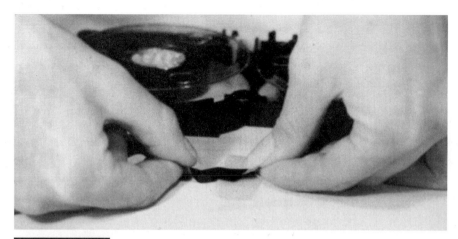

FIGURE 29.12

Aligning the two pieces of video tape.

13. If you did everything right, both ends should be perfectly aligned and appear as a single, unbroken piece of video tape, as shown in Figure 29.13. Press your finger and press down on the video tape to secure it to the clear tape. If the second piece of video tape sticks to the clear tape in the wrong place, don't try to remove it. Cut the section out and start over.

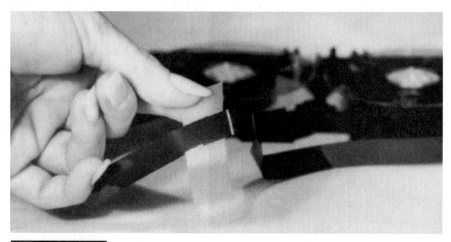

FIGURE 29.13

A perfectly aligned splice.

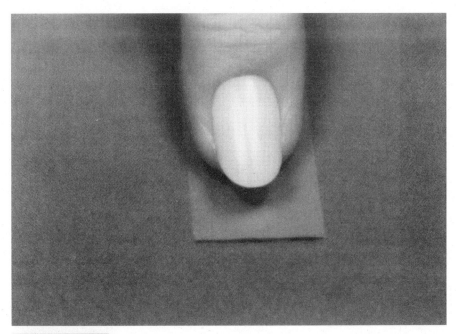

FIGURE 29.14

Leaving a fine line of video tape on the scotch tape.

14. Trim off the excess clear tape.

15. With a sharp pair of scissors, cut a hairline off the edge of the video tape while cutting off the clear tape. Notice the fine line of video tape stuck to the clear tape, as shown in Figure 29.14.

16. Cut the other side the same way. By cutting the clear tape in this manner, you won't leave any exposed adhesive tape to get caught in the video heads.

Testing the Splice

With a clean finger, press down on the front side of the video tape over the splice, as shown in Figure 29.15. If the video tape sticks to you, the sticky part of the clear tape is exposed and will clog the video heads. Cut this section out and start over. You have to be sure that no adhesive is exposed. It might take you two or three times before you get the splice right.

FIGURE 29.15

Testing the splice for exposed scotch tape.

FIGURE 29.16

Pulling the excess tape back into the cassette.

Reassembling the Video Cassette

Turn either reel to pull the excess video tape into the cassette, as shown in Figure 29.16. Now, you are ready to put the video cassette back together.

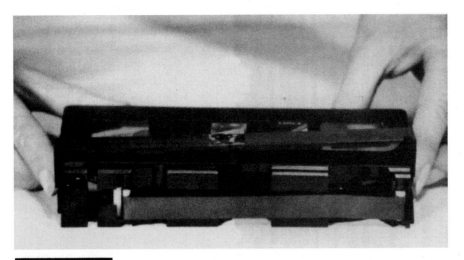

Aligning the video cassette pins.

1. Align both tape-guide pins on the bottom cover to the matching holes on the top cover (see Figure 29.17).

2. Push the two covers together.

3. Flip the cassette over, put the screws back in their proper holes, and tighten each screw.

4. Remove the clear tape to release the cassette door. The video cassette is ready to go.

Review

1. Open the cassette door and tape it back.

2. Take out the five screws located on the back of the cassette.

3. Flip the cassette over and remove the top cover.

4. Be sure to wash your hands before touching the video tape.

5. Release the lock arms to pull out the video tape.

6. Be sure that the video tape is properly threaded through the roller guides.

7. Cut the wrinkled tape off.

8. Cut both ends of the video tape at a slight angle and, at the same time, for a perfect match.

9. Using clear tape, carefully tape both ends of the video tape together.

10. It's important to place the tape only on the backside of the video tape.

11. With a clean finger, check the splice for any exposed adhesive.

12. Turn a reel to pull the excess tape back into the cassette.

13. Reassemble the video cassette.

Repairing Water and Beverage Damage

Liquid is accidentally spilled onto your VCR and some runs down inside the unit. If the liquid lands on any metal or plastic parts, on the transport or, on the cassette carriage itself, the unit will keep on running like nothing happened. If it lands on a circuit board, part or all of the unit will stop functioning. It's important to immediately unplug the VCR, remove the main top cover, and wipe the liquid up from inside the unit. Removing the liquid will keep any metal from rusting. Refer to Chapter 4, "The Cleaning Process."

Water Damage

Water is a conductor of electricity; if water lands on a circuit board, it will short out the board. The board will stop functioning and any function performed by the board will not work. If spilled water causes a board to short out, immediately unplug the VCR, remove the main top cover, and wipe up any excess water with a paper tower. If a lot of water spilled inside, check out the undercarriage as well. After wiping up the water the best as you can, use a blow dryer to completely dry out the unit and the circuit boards.

If you don't have a blow dryer, wait a day or two, then recheck the circuit board to see if the water has completely evaporated. If it is dry, plug the unit in and turn it on. When water contacts a circuit board, the water usually grounds everything out. When the water completely evaporates, the circuitry goes back to normal and the unit starts working again. If the unit doesn't start working, check all the fuses (refer to Chapter 16). Sometimes the water will land in just the right spot and blow a fuse. Replace any blown fuses.

If the circuit board has a lot of dust or lint on it, the water will mix with the dirt, forming a mud-like substance. Because of this substance, it takes a long time for the water to completely evaporate. After the mud-like substance dries, it turns into a mud stain (see Figure 30.1). Clean the circuit board before you plug in the unit.

1. Place the unit on its side and thoroughly spray the mud-stained area with degreaser. The degreaser might not remove the entire stain.

2. If the stain is not completely removed, spray an old toothbrush with degreaser. Brush the stain vigorously, as shown in Figure 30.2.

3. Follow the same procedure for any other stained areas. After scrubbing, re-spray the entire circuit board with degreaser to remove any excess substance left on the board.

In Figure 30.3, the mud stain is on the component side of the circuit board, around and under all the components. Some of these components, such as capacitors and transistors, can be moved out of the way. A capacitor is a round cylinder with two wires attached to the circuit board. A transistor is a small black rectangular box with three wires attached to the circuit board. Usually, these two types of parts are mounted about ¼ inch away from the circuit board on thin wires that bend easily. Gently move these components out of the way, as shown in Figure 30.4.

Other components, such as resistors and ICs, are tightly mounted to the circuit board and can't be moved. A resistor is a long cylinder mounted parallel to the circuit board with the wire coming out of each end. An IC is a black rectangular box with many pins coming out of each side. It has the appearance of a centipede or a postage stamp, and is soldered directly to the circuit board.

FIGURE 30.1

A mud stain on a circuit board.

1. Spray the stained area with degreaser.
2. Use a toothbrush to scrub the circuit board, including the top and around each part, until the board is clean (see Figure 30.5).
3. Re-spray the entire circuit board to remove any substance left on it.
4. Finally, move each part back into its original position, being very careful not to short any wires together or break them off.

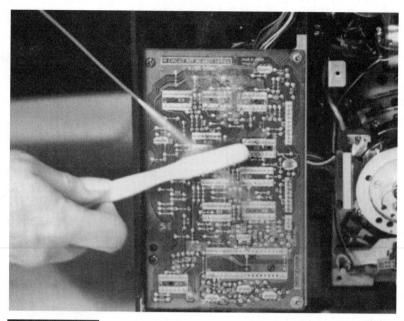

FIGURE 30.2

Scrubbing and spraying with degreaser.

FIGURE 30.3

A stained circuit board on the component side.

FIGURE 30.4

Moving a capacitor out of the way.

FIGURE 30.5

Scrubbing and spraying the component side.

Beverage Damage

If you spill a liquid, such as beer or soda pop into the VCR, the liquid will leave a sticky coating that will take months to dry. Unplug the unit and locate all the stained areas (be sure to check all the nooks and crannies around the affected areas). Clean all affected areas in the VCR, as explained in the previous section of this Chapter. Check for any blown fuses.

Review

1. If any fluids have spilled into the VCR, unplug the unit immediately.

2. Blow dry the unit or let the unit sit for a day or two to allow the liquid to completely evaporate.

3. If the damage is caused by water:
   A. Immediately remove any excess water.
   B. Check to see if the circuit board is clean.
   C. Wait until the water has evaporated.
   D. Plug in the unit. It should work. Replace a fuse, if necessary.

4. If you have a mud-like substance on a circuit board, you'll need to clean it off before you plug in the unit.

5. If you spill a beverage in the VCR, you'll need to clean all affected areas.

6. Check and clean all moving parts that could have come in contact with a beverage.

Repairing a VCR That Has Been Dropped

When you drop the VCR, one of several things can happen.

1. The outside plastic cover cracks or breaks.

2. Some of the mode buttons stop functioning or have broken off. Play, Fast Forward, or Rewind stop working.

3. In the Play mode, the picture is washed out, or there is no picture or sound. The unit also can go completely dead.

4. Another problem a VCR can encounter is improper handling in shipping. The unit was improperly packed in its box and by the time it reaches its destination, it's nonfunctional.

Why a VCR Stops Working

The majority of the time when a VCR has been dropped, a circuit board inside cracks or breaks. A circuit board is the most brittle part

> **QUICK»TIP** Also check for cracks under or around any mounting screw. The mounting screws hold the circuit board in place. If the unit falls on its face, check around all the function buttons and switches. If the unit is completely dead, check the power transformer board.

inside a VCR. The printed circuits break apart, causing the unit to malfunction. Anywhere from one to nine circuit boards are used in a VCR. Most of the circuit boards are located inside the outside cover, surrounding the main chassis or transport. The boards surrounding the main chassis often crack or break on impact.

Locating Cracks

Unplug the unit and locate the point of impact on the outside cover. If the impact was on top, on the side, or on the back, refer to Chapter 2, "Getting inside the VCR." If the impact was on the bottom, refer to Chapter 9, "Getting into the undercarriage." If the impact was on the front, refer to the first three sections in Chapter 32. Locate the circuit board closest to the point of impact.

Between the circuit board and the outside cover are small bumpers made of plastic or rubber that help support the circuit board. They're mounted to the interior of the cover or on the circuit board. On impact, the outside cover pushes the bumper against the circuit board, causing the board to crack or break.

Cracks on Boards Behind the Bottom Plate

After removing the bottom cover plate, check the inside of the cover for bumpers. There are two types of bumpers. One is made of rubber and is mounted to the inside of the cover, as shown in Figure 31.1. The second is made of white plastic and is mounted to the circuit board, as shown in Figure 31.2. In some models, the white bumpers are mounted to the cover, not on the circuit board. If the bumpers are mounted to the cover, check the location between the bumper and the point where it makes contact with the circuit board. If the bumpers are mounted to a circuit board, check around each bumper for cracks. Also, check around each mounting screw and each latch (refer to Figure 31.21).

FIGURE 31.1

A rubber bumper.

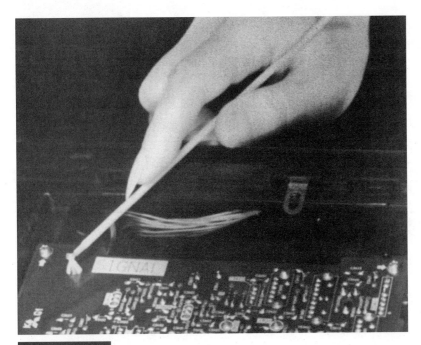

FIGURE 31.2

A white plastic bumper.

Cracks on the Mother Board

1. Remove the main top cover and the transport. Refer to the section in Chapter 9, "Transports."

2. Remove the mother board and you will find plastic bumpers that are mounted to the bottom of the unit (refer to the section in this chapter, "Removing the mother board").

3. Compare the contact point of each bumper to the circuit board. If the VCR is dropped on the bottom, the circuit board will be cracked or broken at these points.

4. Check for cracks around all plugs on the mother board that the transport was plugged into (refer to Figure 9.26).

5. Check around plugs and alignment pins that the head amplifier was plugged into, as shown in Figure 31.3 (I drew arrows pointing at the plugs and the alignment pins).

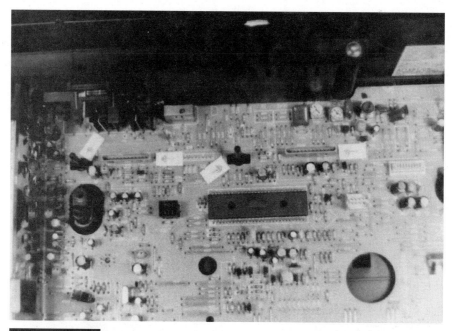

FIGURE 31.3

Video head amplifier plugs and alignment pins attached to the mother board.

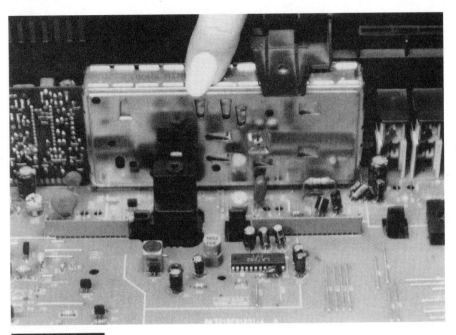

FIGURE 31.4

A tuner inside a metal box mounted to the mother board.

6. Check where the power supply mounts to the mother board (refer to Figure 16.4).

7. Check around the metal box, which is the tuner. It is directly mounted to the board, as shown in Figure 31.4.

Cracks on Boards Behind the Front Cover

1. Remove the front cover and look inside the cover for plastic bumpers, as shown in Figure 31.5. The bumpers might be mounted to the circuit board instead, Figure 31.2.

2. Look inside the cover at the location of each bumper.

3. Compare the contact point of each bumper to the circuit board. If the VCR is dropped on its face, the circuit board will be cracked or broken at these points.

4. If the bumpers are mounted to the circuit board, check around each bumper.

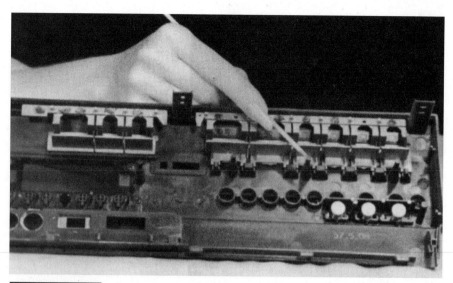

FIGURE 31.5

A plastic bumper.

5. Check around each mode button, slide switch, and check the mounting screws (refer to Figure 32.18 and Figure 32.21).

6. Check around the latches and lips that hold the board in place (see Figures 31.14 and 31.15).

Cracks on Boards Underneath the Top Cover

Remove the main top cover to locate cracks on boards on either side of the VCR. Some circuit boards are mounted by grooves. The board slides into a groove, with a groove on each end of the board. The grooves hold the circuit board in place, as shown in Figure 31.6 (I have pulled the board out of the groove to give you a better view). The grooves differ in length. When a unit falls on its side, the cover pushes against the circuit board and causes the board to bend. The grooves don't bend; consequently, the board breaks right at the groove. If the board is mounted to the chassis with mounting screws, cracks appear around the screws. Check around all latches (see Figure 31.21).

To locate cracks on boards on top of the VCR, look around all mounting screws. If the board has hinges, look around each hinge (Figure 9.8 shows the top hinges). Check around all latches (refer to Figure 3.12).

FIGURE 31.6

A circuit board mounting groove.

Cracks on a Board Mounted to a Power Transformer

If the unit goes completely dead, see if the power transformer is directly connected to a circuit board. Then, check the board for cracks. Figure 31.7 shows where the crack can appear on the circuit board. Directly on the other side of the board is the power transformer. Upon impact, the weight of the power transformer will break the board. Next, check around all the mounting screws for cracks.

In newer models, the power transformer is smaller and is mounted to the mother board (refer to Figure 2.14). Check on the circuit board under the transformer for cracks.

Removing Circuit Boards

The steps required to remove a circuit board varies, depending on the location of the board. The next six sections give instructions for the various locations. Follow the instructions in the section that pertains to your situation.

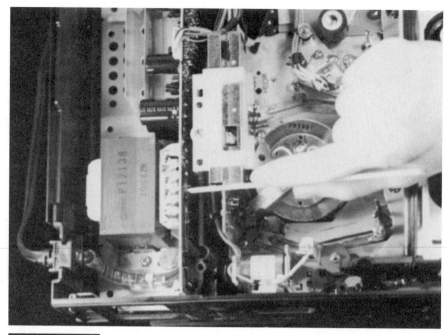

A circuit board connected to a power transformer.

Removing Boards Behind the Bottom Cover

To remove the bottom cover plate and any circuit board under the bottom cover, refer to the section in Chapter 9, "Getting into the undercarriage." To remove any circuit boards attached to the bottom of the transport, refer to the section in Chapter 15, "Removing circuit boards mounted to the transport."

Removing the Mother Board

1. Remove the front cover. Refer to the first three sections in Chapter 32.

2. Remove the transport. Refer to the section in Chapter 9, "Transports."

3. Remove all circuit boards behind the front cover. Refer to the section in this chapter, "Removing boards behind the front cover."

4. If the unit has a power supply that is enclosed with a metal case, to remove the power supply, refer to the section in Chapter 16, "Locating the main power fuse." A plastic panel is attached to all

FIGURE 31.8

The mother board and the rear panel with in and out RCA and RF plugs.

mother boards. This panel contains the video in and out plugs and the RF in and out plugs mounted to it, as shown in Figure 31.8 (I have removed the mother board for a better view).

5. Remove all mounting screws. Two screws might be on top of the rear plastic panel, as shown in Figure 31.9 (I drew arrows pointing at the two screws). One or two screws could be under the VCR, where a half cover plate is located, as shown in Figure 31.10. One or two screws could be on top of the mother board.

6. It might have brackets to hold the top of the tuner and other circuit boards, as shown in Figure 31.11. Remove these brackets.
   A. To remove, push in on the two little clips and slide the brackets straight up and off.
   B. Release all latches above the mother board.
   C. One or two latches are on each side of the board, as shown in Figure 31.12. Push in on the latch above the board, simultaneously pulling that section of the board away from the chassis.
   D. Do the same thing on the other side and remove the board.
   E. A hidden latch might be on the outside, holding the rear panel down, as shown in Figure 31.13. Pull out on the latch, simultaneously pulling the panel up.

FIGURE 31.9

Mounting screws on top of the rear plastic panel.

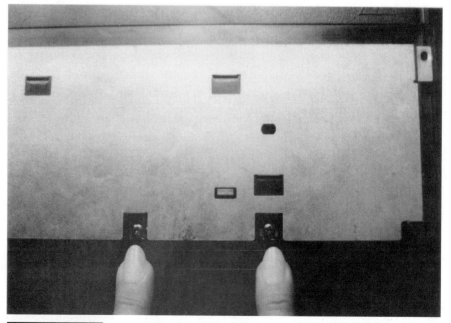

FIGURE 31.10

Mounting screws under the CVR to the mother board.

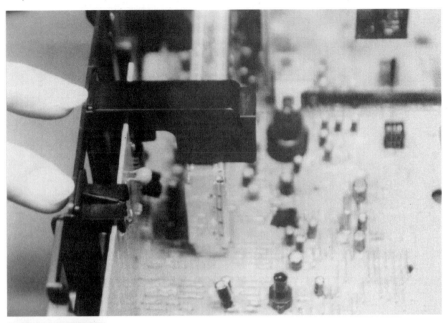

FIGURE 31.11

Brackets holding the top of the tuner and another circuit board.

FIGURE 31.12

Latches on the side of the mother board.

FIGURE 31.13

A hidden latch.

Removing Boards Behind the Front Cover

1. Remove the front cover. Refer to the first three sections in Chapter 32.

2. From one to three circuit boards are mounted to the front of a chassis. These boards might be separated or connected together.

3. In some models, the latches are located on the bottom of the board, as shown in Figure 31.14. There are one to six latches, which vary in size. In some models, these latches protrude through a small square hole in the circuit board. Starting at either end, pull the latch away from the board, simultaneously pulling that section of the board away from the chassis. Open each latch separately in the same manner.

4. After all of the latches have been released, pull the bottom of the board out toward you.

5. Pull the board down to clear each lip across the top of the board, as shown in Figure 31.15.

6. Place the board flat on the table for repairs.

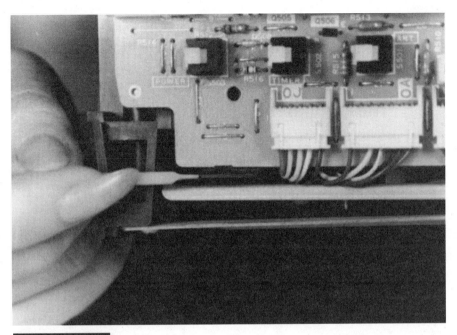

FIGURE 31.14

A latch on the bottom of a circuit board.

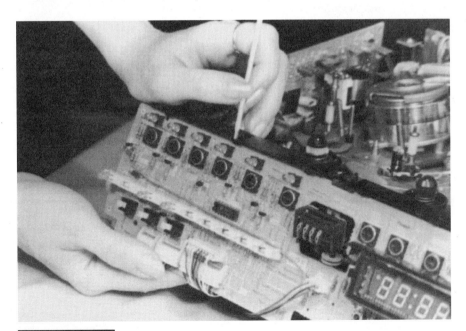

FIGURE 31.15

Pulling down on the circuit board to clear the lips.

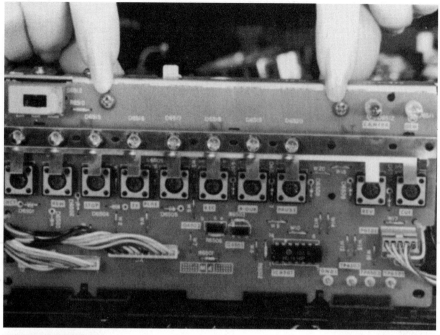

FIGURE 31.16

The circuit boards mounting screws.

In other models, the latches are on top of the board, and the lips on the bottom.

- Simply release the latches on top and pull the top of the circuit board toward you, as previously explained.

- Lift up on the board to clear the lips on the bottom.

- Most boards have extended leads attached to the circuit board. You do not need to unplug the circuit board, just stretch the leads out and lay the circuit board onto the table for repair.

Another model has screws across the top or bottom of the circuit board, as shown in Figure 31.16.

- Remove the screws and slide the board in the direction of the screws, clearing the groove across the opposite end of the board.

- Lay the board flat on its face for repairs.

FIGURE 31.17

A latch holding the top of the front circuit board.

A different model has mounting screws completely around the circuit board.

■ Remove the screws and pull the board straight off.

■ Lay the board flat on its face for repairs.

In another model, the circuit boards on the front of the chassis are connected to the circuit board on the bottom of the chassis. Refer to the sections in Chapter 9, "Fifth way" under the subhead "Way to open a circuit board."

In newer models, the front circuit boards plug into the mother board. These boards vary in size and shape. They have from one to four latches, which vary in size, at the top of the circuit board, as shown in Figure 31.17.

■ Pull the latch away from the board, simultaneously pulling that section of the board away from the chassis. Do each latch separately until the board is disconnected from the chassis.

■ Lift the board straight up and the plugs will disconnect from the mother board (see Figure 31.18).

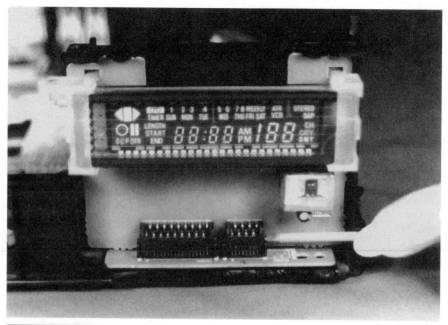

Location of the plugs under the front circuit board.

In other newer models, instead of the front circuit board plugging directly into the mother board, there are ribbons, as shown in Figure 31.19.

- Release the latches and flip the board over.
- After pulling the circuit board out, look at the chassis. You'll find plastic bumpers supporting the back of the circuit board, as shown in Figure 31.20. These bumpers also can cause the board to crack.
- Check the circuit board at these locations. Some units have two sets of bumpers, one in front of and one behind each circuit board. In all other units, there will be bumpers on one side, or no bumpers at all.

Removing Boards on Either Side

There are three ways to remove boards on the sides of your VCR.

1. Pull the board straight up and out of the groove on the bottom of each side (refer back to Figure 31.6).

2. If the boards are mounted to the chassis by mounting screws, remove the screws and pull off the board.

FIGURE 31.19

A ribbon attached to the front circuit board.

FIGURE 31.20

Plastic bumpers behind the circuit board.

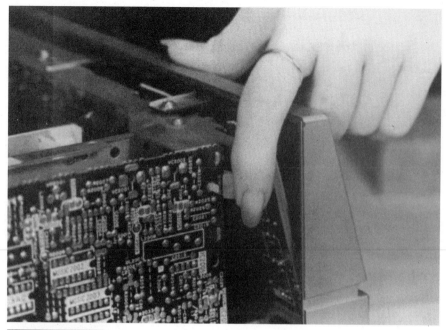

FIGURE 31.21

Releasing a white latch.

In other models, a white latch with a pin is protruding through a hole in the board near the top, with one latch on each side of the circuit board. Pull the latch back while pulling the board toward you, as shown in Figure 31.21. Repeat this procedure for the other end. The board opens like a door.

Removing Boards Underneath the Top Cover

For boards covering the top of the transport, refer to the section in Chapter 3, "Removing circuit boards blocking the tape path." When removing boards located on top and to the right side of the transport, you will find 4 to 6 mounting screws which are red or gold in color. It might have arrows pointing at each mounting screw (refer to Figure 9.2) and black plastic clips around the edges (refer to Figure 3.12). In some units, a black plastic panel on the back of the unit comes up with the board (refer to Figure 3.14). This panel has the video in and out plugs and the RF in and out plugs mounted to it.

In a few models, in addition to the mounting screws, a double white clip is in the front right corner, inserted through a square hole on the board. Squeeze the white clip together and pull up the board.

In one or two models, the leads coming from the video head hold the circuit board down. Look on the component side of the board for a metal box with a snap-on lid. Pop the lid off and disconnect the plug inside.

Removing the Board Connected to a Power Transformer

Before removing a board connected to a power transformer (see Figure 31.7), you must unplug the VCR. Electrical shock is possible. There are two ways to remove this type of board. In both cases, the power transformer comes out with the circuit board. The first type has two Phillips head mounting screws at the base of the transformer that are screwed into the plastic part of the chassis. Remove these screws. Pull the power transformer and the circuit board straight out.

The second type of board has the power transformer and the circuit board mounted to a metal shield under the transformer.

1. Remove the mounting screws from the shield. It has three or four mounting screws, usually in each corner of the shield. Sometimes a screw is in the middle.

2. Two screws are at the base of the transformer. These two screws hold the transformer to the metal shield. The mounting screws with arrows are pointed out in Figure 31.22. Remove the screws and pick up the entire power supply, as shown in Figure 31.23. Some screws have black grounding wires attached.

3. From one to three white plugs are attached to this board. To make the repairs easier, detach the plugs. Each plug has a different number of pins. When replacing the board, you'll have no difficulty matching which plug goes where. The AC cord is attached to this board and it comes up with it.

In some models, one screw is mounted to the bottom cover plate, under the power supply. Take out the screw to remove the power supply.

Another model has four screws at the base of the transformer with black grounding wires attached to the screws. Remove all screws and grounding wires.

In newer models, the power transformer is much smaller and is mounted to the mother board. You must remove the mother board to check under the transformer. Refer to the section in this chapter,

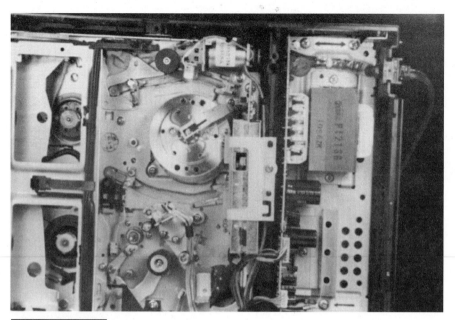

FIGURE 31.22

Arrows pointing at the mounting screws.

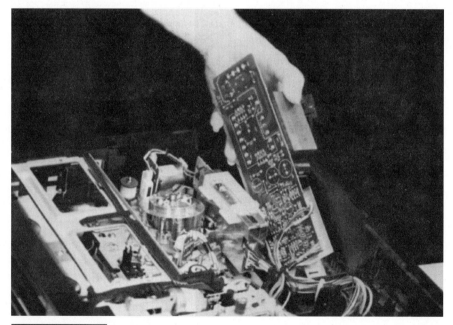

FIGURE 31.23

Removing the power supply.

"Removing the mother board." In other models, the power supply is in an enclosed metal case. Refer to the section in Chapter 16, "Locating the main power fuse."

Making the Repairs

The following sections will show you how to repair cracks to any of the circuit boards in your VCR. Read each of the sections before you begin, then carefully follow the instructions.

Repairing Cracks on Circuit Boards

The component side of a circuit board usually contains a fine white line or branches of lines, as shown in Figure 31.24. These lines can be difficult to detect. If the board is broken, the crack will be obvious and the board probably will be broken, as shown in Figure 31.25. Realign the break in the board so that it's flush, leaving just a crack in the board. On the component side only, run a bead of super glue along the full length of the crack to prevent it from spreading, as shown in Figure 31.26. Hold the board together until the glue sets.

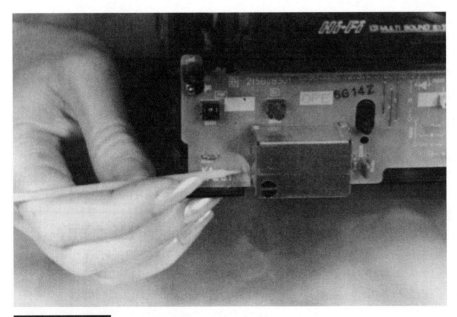

FIGURE 31.24

A fine white crack on the component side of a circuit board.

FIGURE 31.25

A broken circuit board.

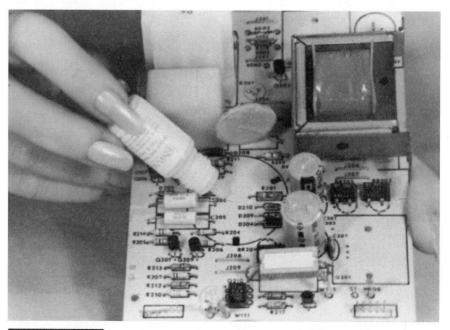

FIGURE 31.26

Mending a board with glue on the component side.

FIGURE 31.27

A cracked circuit board.

Figure 31.27 shows how a crack appears on the opposite side, the circuitry side, of the board. Look along this crack. Notice the points where the circuitry tracks have been broken and no longer make contact. This interruption has caused certain functions to stop working. To repair these broken tracks, solder a wire across the broken track.

Repairing a Broken Track

Sometimes the cracks branch out, as shown in Figure 31.28. The branches are hairline cracks and are difficult to see. Be sure to locate the entire crack and all the branches.

1. Starting at one end, follow the cracks until you come across the first broken circuit track. Tracks are part of the circuitry and connect the components together. The tracks are bonded to the bottom of the board and are like thin tin foil. They are covered with green- or brown-colored plastic coating, which is an insulation.

To solder a wire to a circuit board:

1. Remove about ⅛ inch of insulation from a #24 stranded wire, then coat the end of the wire with a thin layer of solder.

2. Place the coated wire on top of a connection (a connection is where a lead from a component, such as a resistor, comes through a little hole in the circuit board and is soldered to the circuitry side of the board and a track is connected to the same connection).

3. Put the hot solder iron on top of the wire and push a small amount of solder onto the connection.

4. Remove the solder, leaving the solder iron on the connection for a second or two. It's important not to let the solder run into neighboring connections. If you do, the circuitry will short out. If you get solder on other connections, use solder wick to remove the solder. Refer to the section in Chapter 17, "Using solder wick to remove old solder."

5. Follow the same procedure for soldering all wires to circuit boards.

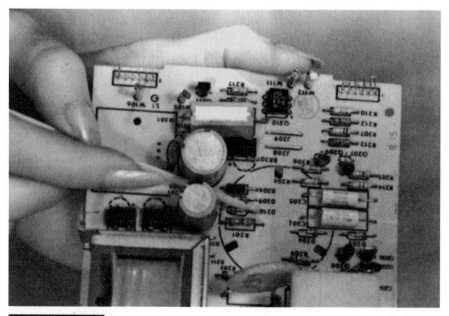

FIGURE 31.28

The crack branches out.

FIGURE 31.29

Pointing at the crack and its closest soldering connection.

2. As soon as you locate the first broken track, follow the track until you come to the first solder connection.

3. Solder the solder coated wire to this connection. Figure 31.29 points to the track at the break and the wire is soldered to the first connection on the same track.

4. Retrace the track back to the break and continue on the other side of the break until you come to its first soldering connection.

5. Take the wire you soldered to the first connection and stretch it tightly across to the second connection, adding approximately ⅛ inch of wire.

6. Cut and remove the insulation from the wire.

7. Coat the end of the wire with a thin layer of solder.

8. Solder that end of the wire to that soldering connection.

9. Retrace the track to be sure that the wire is soldered to the same track on both sides of the crack (Figure 31.30). The tracks are close together, so it's easy to get them confused. There can be a long distance between soldering points. Look at Figure 31.31. Notice how far away the white wire is from the track, where indicated by the pointer, and the odd location of the soldering connections found on each side of the crack.

Repairing Double-Printed Circuit Boards

A few models have double-printed circuit boards. The circuitry side of the board looks the same as previously shown, but the component side also has tracks around the components. On this type of board, you must repair the tracks on both sides of the board. It's important to stay on the right track. Crossing a track can cause more damage to the unit. Recheck all the tracks before plugging in and turning on the unit.

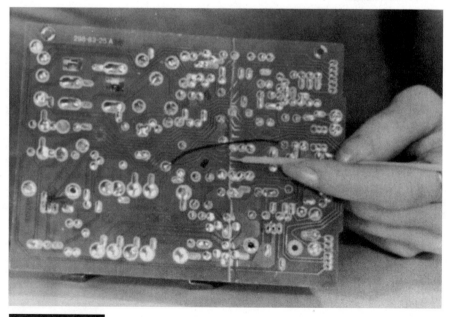

FIGURE 31.30

The closest connection on the other side of the crack.

The longest distance to the closest soldering connections.

Repairing Broken Covers

The easiest way to repair a broken cover is with super glue. Take a tube of super glue and run a bead of glue down the crack or the broken piece on the inside of the cover only. Hold the two pieces together until the glue sets.

Review

1. Look for the impact point.

2. Open the unit and look for the circuit board closest to the impact point.

3. Remove the circuit board and look for a crack that can be found around the mounting screws or where a bumper sits up against a board.

4. If the board is cocked, straighten the board out, making it flush.

5. No matter if you have a crack or a break, glue it on the component side only.

6. Trace the broken tracks carefully and solder a wire to each end of the broken track.

7. Do not cross the tracks.

8. Be careful not to solder two neighboring connections together and cause a short.

9. Remember to retrace all connections first before plugging in the unit.

10. If the unit goes completely dead, check around the power transformer for a cracked circuit board.

Repairing Broken or Nonfunctional Buttons

The VCR was accidentally bumped or dropped on the front cover. A function button has broken off, is loose, or no longer functions. You would assume that each button is a separate piece, but most buttons are manufactured right into the front cover. A VCR repair shop would probably replace the entire front panel cover, which is expensive and unnecessary.

First Steps in Removing a Front Panel

The main top cover covers the mounting screws and the latches to the front cover. Remove the main top cover. Refer to Chapter 2, "Getting inside the VCR." The top of the front cover is mounted in one of three ways.

1. Two to four Phillips head-mounting screws run across the lip on top of the front cover.

2. Three to four locking latches are placed across the top of the front cover with a screw in each one, as shown in Figure 32.1.

FIGURE 32.1

A locking latch with its screw inserted.

3. Three to four locking latches are located across the top of the front cover with no screws, as shown in Figure 32.2, or the latches can be barely sticking out from under the top of the front cover, as shown in Figure 32.3 (I have removed the front cover so you have a better view of the latches).

A bracket might be located across the top latches.

TRADE SECRET In some models, the bottom cover plate covers the bottom latches on the front cover and the mounting screws to the bottom cover go through the latches. To remove the bottom cover, refer to the section in Chapter 9, "Removing the bottom cover plate." Other units have no bottom cover, only latches. In all models, these are the first steps to follow.

1. To remove this bracket, refer to the section in Chapter 8, "Preparation for removing cassette carriages."

2. Remove any screws on top of the front cover.

3. Open any panel doors located on the front of the unit.

4. Some models have a hidden screw inside a door. If so, remove it.

5. Lay the unit on its side.

FIGURE 32.2

A locking latch with no screw.

FIGURE 32.3

A latch sticking out under the top of the front cover.

Removing Front Covers

Look for locking latches on the top, ends, and the bottom of the front cover (I have removed a front cover to show all the latches around the outside of the cover, as shown in Figure 32.4). These latches have a round plastic pin protruding through a hole in the latch, as shown in Figure 32.5. In other models, the latches have a rectangular-shaped clip protruding through a hole in the latch (see Figure 32.2).

To release the latches, pull it out until it clears the pin or clip protruding through the hole. Pull out on that section of the front cover to keep the latch from recatching, as shown in Figure 32.6. To release the front cover, read each of the following sections and follow the instructions that pertains to your unit.

First Type

In models with three latches on top and a latch on each end:

1. Release the right top corner latch and the latch on the right side.

2. One at a time, go back and release the latch on top in the middle, the latch on the top left side, the latch on the left side, and all the bottom latches.

3. Remove the front cover.

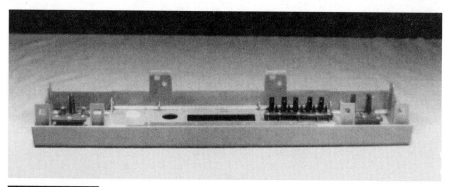

FIGURE 32.4

All the latches around the outside of the front cover.

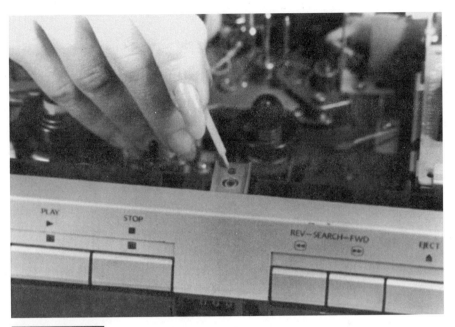

FIGURE 32.5

A plastic pin protruding through the latch.

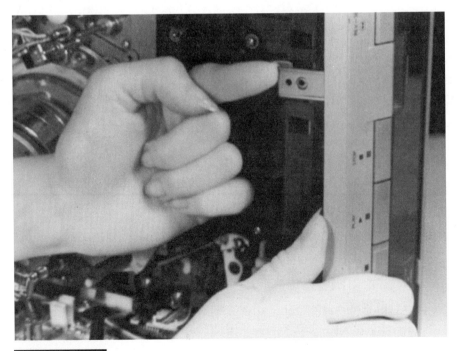

FIGURE 32.6

Pulling out on the cover to keep the latch from re-latching.

Second Type

In models with three latches on top and no side latches:

1. Release the top corner latch, the middle latch, the other top corner latch, and then all the bottom latches, one at a time.

2. Remove the front cover.

Third Type

In models with four latches on top and a latch on each end:

1. Release the right two top corner latches and the latch on the right side.

2. Release the two latches on the top left side, the latch on the left side, and then all the bottom latches, one at a time.

3. Remove the front cover.

Fourth Type

Models with three to four latches on top, no side latches, and no visible latches on the bottom, have three to four hidden latches. To remove this type of front cover, release the top latches, pull the top of the front cover forward and release the bottom latches. Figure. 32.7 shows a hidden latch. Each hidden latch consists of a slot on the chassis and a hook on the front cover. The hook slips into a slot to hold on the bottom of the cover. In other models, the hook is on the chassis and the slot is on the bottom of the front cover. The principle, however, remains the same.

Fifth Type

Older models have no latches. The front cover is mounted with two to six Phillips head screws. The two locations for these mounting screws are across the top and bottom of the front cover or on each end of the front cover. Just remove the mounting screws and pull the front cover straight off.

Sixth Type

In a few models, after removing the front cover, you'll find a black wire attached to it. This black wire grounds the middle of the front

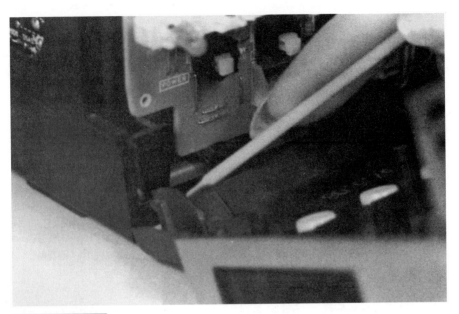

FIGURE 32.7

A hidden latch.

cover to the chassis. The wire is always attached to the front cover by a small plug. Simply unplug the wire from the cover. Most models don't have a grounding wire. Just be aware of the possibility.

Remounting Front Covers

If your unit has a cassette door mounted to the front cover:

1. Locate the automatic door opener lever, which is located usually on the front right side of the cassette carriage, as shown in Figure 32.8.

2. Position the pin on the door lever so that it goes in front of the tab located on the right side of the cassette door, as shown in Figure 32.9 (I have removed the cassette door to give a better view of the tab).

3. The easiest way to do this is to open the door all the way up and hold it there until the front cover is in place.

4. Align any controls or switch knobs through the proper hole in the front cover.

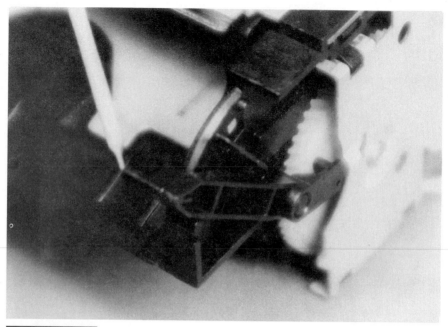

FIGURE 32.8

The pin on the door opener lever.

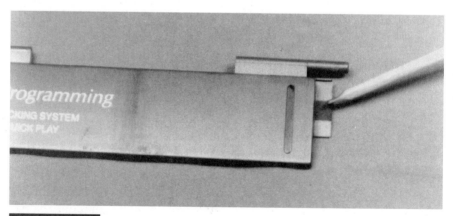

FIGURE 32.9

The tab located on the right side of the cassette door.

5. If the unit only has mounting screws:
   A. Push the cover straight on and then release the cassette door.
   B. Insert and tighten the mounting screws.

6. If the unit has latches on the top, bottom, and sides:
   A. Push the cover straight on until each latch pops back into place and then release the cassette door.
   B. Insert and tighten any mounting screws.

7. If the cover has hidden latches:
   A. Insert the hook into the slot at the bottom of the front cover and push the top cover shut like a door.
   B. Continue pushing until each latch pops into place.
   C. Insert and tighten any mounting screws.

Locating and Repairing Missing Buttons

Most buttons have a lip around their base to keep the button from falling out. The missing button usually is floating around inside of the front cover.

Buttons break off at their hinge. The hinge usually is made of two plastic fingers at the base of the button, as shown in Figure 32.10. In other types, the two hinges curve around the button, as shown in Figure 32.11, or the hinge could be one solid piece, as shown in Figure 32.12. Place the front cover on its face. Reposition the button back through its proper hole on the cover. Then, align the broken fingers on the hinge.

Two methods are used for repairing buttons. I highly recommend using the clear silicone glue method. Apply clear silicone glue to the hinge. Completely cover both fingers of the hinge with the silicone (see Figure 32.10). The silicone glues the hinge, but remains flexible. If the hinge is partially or completely missing, use the silicone glue to form a new one. Use the silicone liberally. The bonding is what's important, not the appearance.

Another method for repair is to place a small drop of super glue onto the break of the hinge, as shown in Figure 32.13. Hold the button in place until the glue is set. Use a piece of fiber tape or black electrical tape to hold it into position permanently, as shown in Figure 32.14. The hinges on the button must remain flexible.

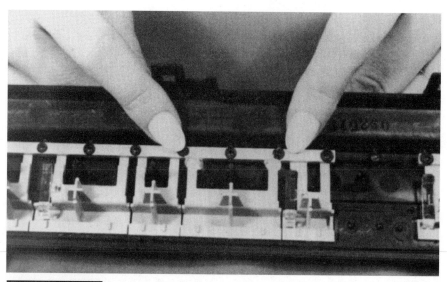

FIGURE 32.10

The two hinges repaired with silicone glue.

FIGURE 32.11

U-shaped hinges.

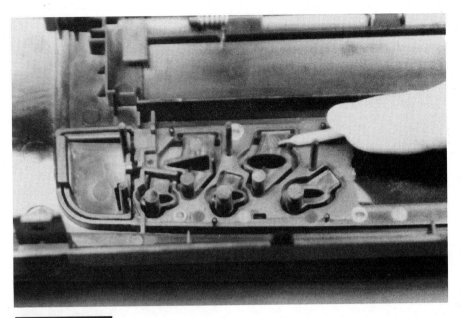

FIGURE 32.12

A solid flexible hinge.

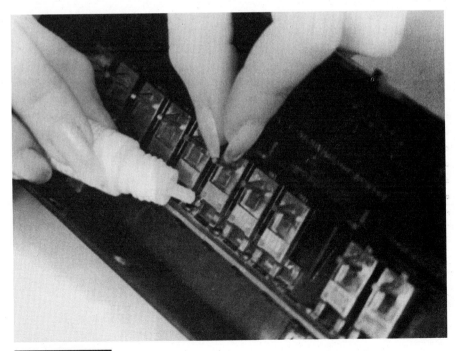

FIGURE 32.13

Repairing the broken fingers on a hinge.

Taping a button to hold it into position.

Sometimes a button is held on with a shaft instead of a hinge. This shaft breaks right at the base of the button. The button falls outside of the VCR, instead of the inside, leaving the shaft protruding through the button hole, as shown in Figure 32.15. Drop a bead of super glue onto the end of the shaft, place the button up against the shaft, and hold them together until the glue is set.

Repairing Twisted Buttons

If a function button is cocked or out of alignment, one of the fingers is broken on the hinge. Twist the button back into position and place a drop of super or silicone glue onto the broken finger. Hold it in place until it's set.

FIGURE 32.15

A shaft protruding through the button hole.

Why Mode Buttons Become Nonfunctional

A button is nonfunctional if it's intact, but doesn't function. Three different problems can cause this to happen. First, the shaft behind the button has broken off, as shown in Figure 32.16. If you can find the broken piece of shaft, simply super glue it back into place. If you can't find the piece of shaft, follow this procedure: Remove the empty ink well from an ink pin and cut a section off the same length as the other button shafts. Then, super glue it to the broken shaft.

The second cause is a broken micro switch. Each button mounted to the front cover has a shaft protruding out the back of the button, as shown in Figure 32.17. When pushing on a button, the shaft on the back of the button pushes against a micro switch. The micro switches are located directly behind each button on a circuit board, as shown in Figure 32.18. These micro switches can fall apart on impact. Although the micro switch has been broken, the button on the front cover might be intact.

Incorporated in the center of the micro switch is a little plastic button. On impact, the button pops out, leaving the switch nonfunctional,

FIGURE 32.16

A broken off button shaft.

FIGURE 32.17

The shaft protruding out the back of the buttons.

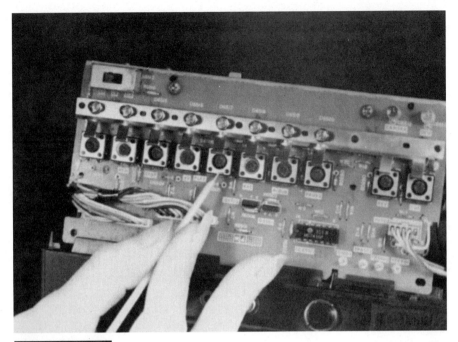

FIGURE 32.18

A micro-switch on a circuit board.

as shown in Figure 32.19. In this case, the micro switch has to be replaced.

The third cause is a cracked circuit board. Refer to Chapter 31, "Repairing a VCR that has been dropped."

Replacing Micro Switches

1. Unplug the unit and remove the circuit board where the micro switch is mounted. Refer to the sections in Chapter 31, "Removing boards behind the front cover" under the subhead "Removing circuit boards."

2. After removing the circuit board, turn it over.

3. With the backside facing you, locate the broken micro switch and the pins protruding through the back of the board. To aid in identification, a black or white square outline is around each set of micro switch pins, as shown in Figure 32.20. You'll find four or five pins within each square.

FIGURE 32.19

The missing button on a micro-switch.

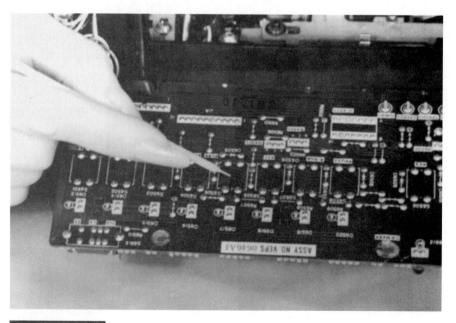

FIGURE 32.20

The special white marking indicating each micro-switch.

4. Where the pins come through the board, place solder wick over one of the soldered connections. Refer to the section in Chapter 17, "Using solder wick to remove old solder."

5. After removing the solder, turn the board over and pull out the micro switch. Refer to Chapter 1, "Parts," for replacement.

6. To replace the micro switch, place the pins on the micro switch through the holes on the circuit board. The pins on the micro switch are offset, so a switch can't be put in backwards.

7. Turn the board over. Refer to the section in Chapter 17, "Soldering pins to a circuit board."

8. Be careful not to solder any pins together. If you do, use solder wick to remove any excess solder.

Replacing Slide Switches

A slide switch (see Figure 32.21) also stops functioning upon impact. Slide switches are used for setting the tape speed, stereo hi-fi, audio inputs, and so on. On impact, the switches fall apart. Remove the cir-

FIGURE 32.21

A slide switch.

cuit board attached to the switch. To replace these switches, use the same method as previously explained for micro switches. The only difference is the number of pins. These switches have anywhere from 4 to 12 pins per switch. The pins have a rectangular outline around them, as shown in the lower left corner of Figure 32.20. Inside this outline is where the pins for the slide switch are located.

Review

1. Remove the main top cover.
2. Remove the bottom cover plate.
3. Remove any screws mounted to the latches.
4. Check all panel doors for a hidden screw.
5. Release all latches around the front cover.
6. If there are hidden latches, pull the top of the front cover toward you, and the bottom will release.
7. Properly position the broken button.
8. Use silicone glue to repair any parts on a broken hinge.
9. The hinge must remain flexible.
10. If a button is nonfunctional, check for a broken micro switch or a cracked circuit board.
11. Remove all switches by unsoldering their pins.
12. Be sure not to solder two pins together when replacing a switch.

Repairing Remote Controls

Three common problems occur in remote controls.

1. The batteries get weak or go bad.

2. If it has been dropped, the circuit board inside can crack or break.

3. If a beverage has been spilled on it, the buttons will stick down and some or all of the functions stop working.

If replacing the batteries doesn't correct the problem, take the remote apart for repairs.

How the Remote Works

The main active part of a remote is called an *infrared LED (light-emitting diode).* An infrared LED is an invisible light of wavelengths longer than 750 nanometers, characterized in terms of radiometric quantities. All remote controls have an infrared device. These devices look like small light bulbs. These LEDs send out a pulsating infrared

FIGURE 33.1

LED protruding through the front of the unit.

light, which signals the VCR what function to perform. Some units have three LEDs, but most have one or two. They face toward the front from behind a red plastic window, which is usually connected to the back cover. Other remotes have no plastic window and the LED protrudes directly through the front, as shown in Figure 33.1.

Locating the Mounting Screws

1. Remove the battery cover door on the back of the remote and take out the batteries. There are four locations where the mounting screws might be. All mounting screws are on the back of the remote. The first type has one small Phillips head-mounting screw in the middle, above the battery compartment, as shown in Figure 33.2. The second type has two Phillips head screws inside the battery compartment, as shown in Figure 33.3. The third type has the mounting screws hidden under rubber foot pads. In most cases, the two pads opposite the battery compartment contain the mounting screws.

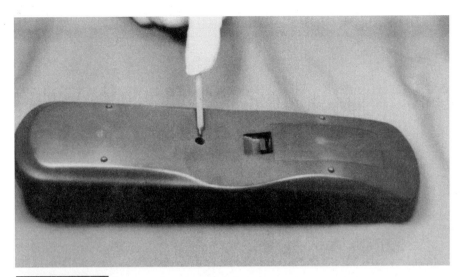

FIGURE 33.2

The back cover mounting screw.

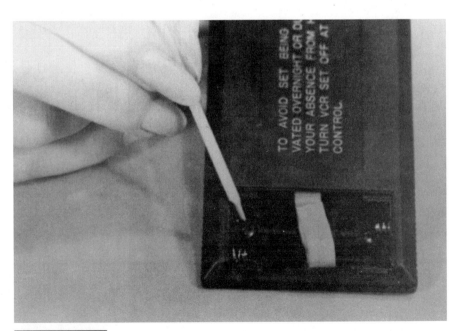

FIGURE 33.3

Two mounting screws in the battery compartment.

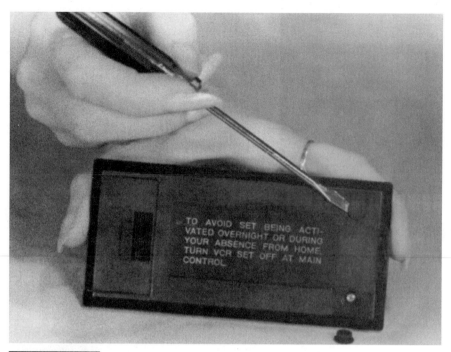

FIGURE 33.4

Lifting the pad to locate the mounting screws.

2. With a small flathead screwdriver, pop out the rubber foot pad, as shown in Figure 33.4. The last type has four Phillips head screws, with one in each corner, as shown in Figure 33.5.

3. Remove all the mounting screws.

Removing the Back Cover

1. Look at the side of the remote where the top and bottom covers come together.

2. If a small indentation is on each side of the remote (see Figure 33.6), follow this next procedure. After removing the mounting screws from the battery compartment, lift that end of the remote up approximately ¼ inch.

   A. Push a small flathead screwdriver into the indentation to release the latch. The back cover will pop up approximately ⅛ inch, as shown in Figure 33.6.

   B. Turn the remote around and release the other latch in the same manner.

   C. Pull the back cover of the remote up to release the front latches, as shown in Figure 33.7.

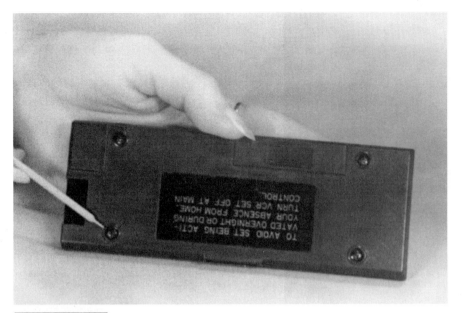

FIGURE 33.5

Mounting screws in each corner.

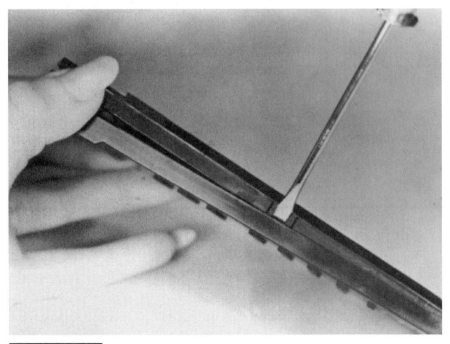

FIGURE 33.6

Releasing the latch.

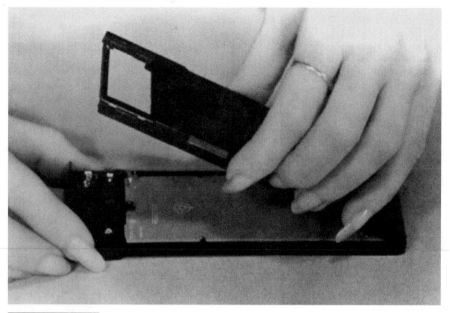

FIGURE 33.7

Releasing the front latches.

3. If no indentations are on either side of the remote, follow this procedure. Start by removing the mounting screws.

In older models:

1. Slide the bottom cover straight back toward the battery compartment approximately ¼ inch to release all the latches, as shown in Figure 33-8.

2. The cover then lifts right off.

In newer models:

1. Slide the bottom cover toward the battery compartment, which is in the front of the remote, approximately ¼ inch to release all the latches. You might need to pull a little hard before it releases. If it doesn't release, read the next section in this chapter.

2. In some remotes, be careful with the battery connections after it releases because these connections are made of a flexible spring wire and you could break them off or bend them.

3. Just slide the springs through the spring hole in the battery compartment, as shown in Figure 33.9, and lift the back cover off.

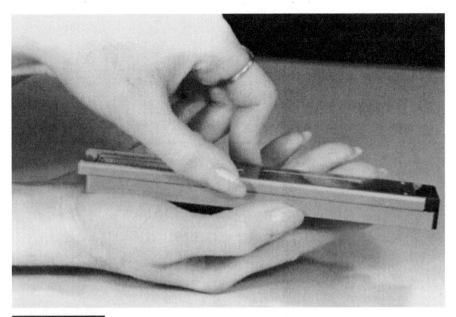

FIGURE 33.8

Pulling straight back to release all the latches.

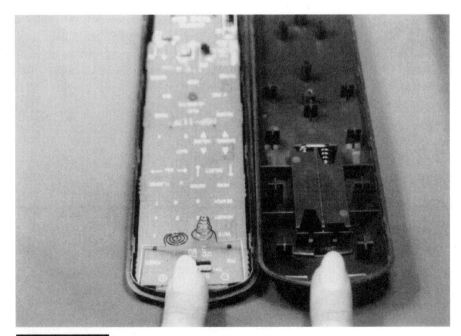

FIGURE 33.9

The spring on the circuit board and the spring holes on the back cover.

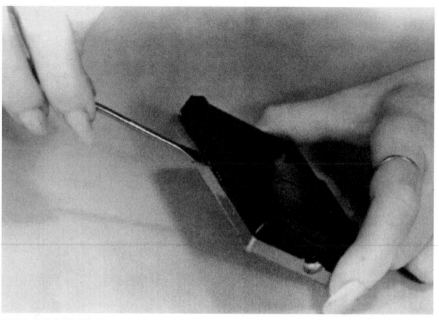

FIGURE 33.10

Popping off the back cover.

Another type of remote has no mounting screws or indentations.

1. Shove the tip of a small, flathead screwdriver between the top and bottom cover.

2. Twist the screwdriver and that section will pop up, as shown in Figure 33.10.

In newer models, look close in the groove where the two covers are connected together and you might see a very small indentation showing where each latch is. Two to four latches should be on each side and one to two latches should be on each end. Follow the same procedure all the way around the remote. Now you can lift off the back cover.

The last type has a mounting screw located in each corner of the remote control. Simply remove the screws and lift off the back.

Removing the Window

If the window is in the way when removing the circuit board, and remains with the top portion of the remote control, place a small flathead screwdriver at the base of the window and slide the window up and off as shown in Figure 33.11.

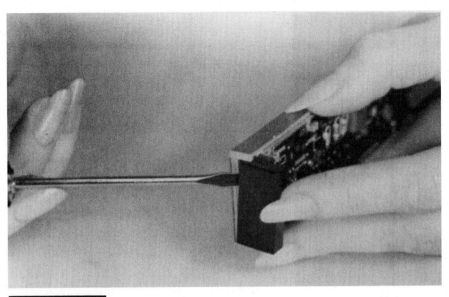

FIGURE 33.11

Removing the window.

Removing All Circuit Boards

After removing the circuit board, don't flip the front cover over because some remotes have separate buttons. If you flip it over, the buttons will fall out and you will get them mixed up.

In the first type of remote control, you'll notice that two battery terminals are attached to the circuit board where the batteries come in contact. With a small flathead screwdriver, pry the battery terminals straight up, as shown in Figure 33.12. The circuit board now will lift straight up and off.

In other models, the circuit board has notches on each side. Some of these notches have plastic latches that hold the circuit board in place. Starting at the back of the board, simply place the tip of your fingernail underneath the circuit board while prying it up. At the same time, use your other hand to push the latch back, releasing the circuit board, as shown in Figure 33.13. Follow the same procedure for each latch. After releasing all the latches, the circuit board will lift straight up and off.

In older models, the circuit boards have from one to five small Phillips head mounting screws. Remove the screws and lift the board straight up and off.

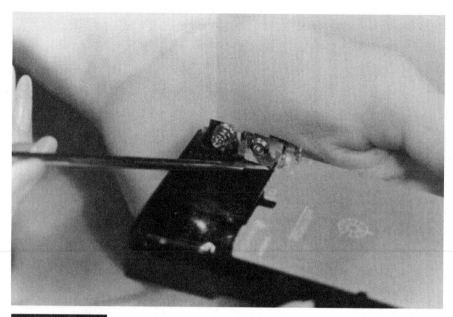

FIGURE 33.12

Prying up the battery terminals.

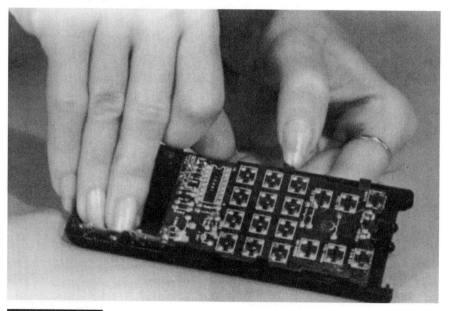

FIGURE 33.13

Pulling back the latch to release the circuit board.

FIGURE 33.14

The wires attached to the battery compartment.

In some older models, after removing the circuit board from the front cover, you'll find a red and black wire attached to the circuit board and the battery compartment, as shown in Figure 33.14. Don't move these wires more than necessary; they can break off.

In some newer models, a knob is on front of the remote that holds the circuit board in place. Place a small flathead screwdriver under the knob and pry the knob off, as shown in Figure 33.15. Then, lift the circuit board straight up and off.

In other newer models, the circuit board is attached to the back cover by the battery springs. To remove the board, slide the battery springs on the circuit board out through the spring holes in the middle of the back cover. Then, lift the board straight up and off. Figure 33.16 shows the spring holes in the middle of the back cover (left), the circuit board with the battery springs (center), and the front cover with the rubber pad (right).

FIGURE 33.15

Prying up on the control knob.

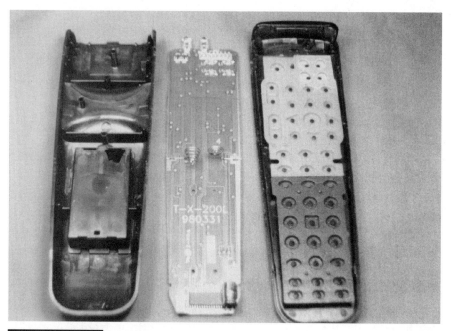

FIGURE 33.16

The back cover with the spring holes (left), circuit board with battery springs (center), front cover with rubber pad (right).

FIGURE 33.17

The back view of the function buttons.

Repairing Buttons That Stick Down

If a beverage has been spilled on the face of the remote, the buttons will stick and the remote will stop working. Follow this procedure.

In older models:

1. When the circuit board is removed, you'll find a rubber pad. This pad usually comes off with the circuit board while the buttons remain in the front cover, as shown in Figure 33.17. Place a hand on the back of the front cover and turn the cover on its side so that the buttons won't fall out.

2. Make note of the location of any colored buttons, so you'll be able to replace the colored buttons back in their proper holes.

3. Flip the cover over and let the buttons fall out onto the working surface.

4. Spray the front cover and use a toothbrush with degreaser, or you can use water.

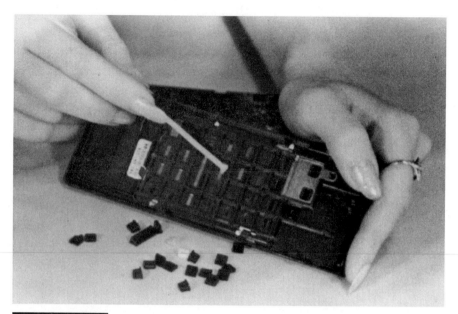

FIGURE 33.18

Cleaning button holes with a glass brush.

5. Brush the entire cover, especially down in each hole.

6. Flip the top cover over and clean the other side in the same manner.

7. Then, respray the cover with degreaser or rinse it with water.

8. With a glass brush, clean each button hole in the front cover, as shown in Figure 33.18.

9. Spray some degreaser onto your glass brush or you can use water and scrub each side of each button until it is clean, as shown in Figure 33.19.

10. Remember to periodically clean the brush.

11. After scrubbing each button clean, be sure to respray them or rinse them off and then dry.

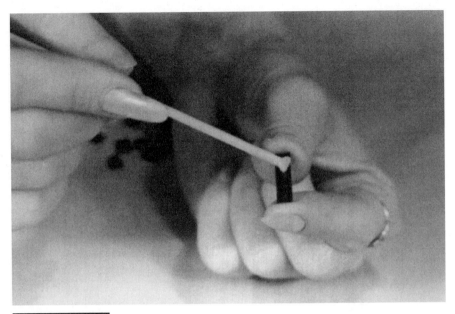

FIGURE 33.19

Cleaning the buttons.

In newer units:

1. A rubber pad has buttons incorporated into the pad. Simply pick the pad up (Figure 33.20) and follow the same procedure as previously explained for cleaning the top cover and button holes.

2. Spray or use water and brush the top of the pad where the buttons are until it is clean.

3. Be sure to clean around the edges of each button on the pad and the bottom holes in the front cover, then dry.

Repairing Nonfunctional Buttons

The main reason that a function button stops working is a buildup of dirt or a dried beverage on the contact of each button. When the button is pushed, it fails to make electrical contact and the button doesn't work.

FIGURE 33.20

Removing the pad.

To fix this problem:

1. Turn over the pad that you just cleaned.

2. On the back of the pad are round- or rectangular-shaped conductors. Dip a glass brush into some cleaning alcohol.

3. Use the brush to clean all the conductors, as shown in Figure 33.21.

4. After cleaning each contact, reclean the brush with a paper towel.

5. After cleaning all the conductors, spray the entire pad off with degreaser to remove any fibers left by the glass brush.

6. On the back of each circuit board are two types of printed contacts (Figure 33.22), which look like small and large S's. These printed contacts come in contact with the conductors on the pad. To clean these contacts, use a glass brush to clean each contact as you did with the conductors.

7. After cleaning each contact, spray them with degreaser to rinse them off.

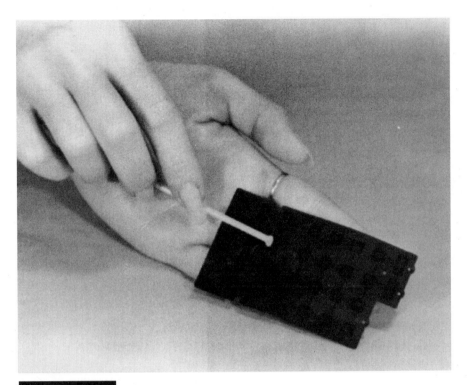

FIGURE 33.21

Cleaning the conductors.

FIGURE 33.22

The two types of printed contacts.

In older units where the buttons are separate from the pad, you should find a rubber pad stuck to the face of the circuit board. Pull it off. Clean the conductors on the back of the pad and all the printed contacts on the circuit board, as previously explained. This repairs all non-functioning buttons.

Repairing Dropped Remotes

When a remote is dropped, the circuit board might break or crack, causing the remote to stop working. To repair a cracked or broken circuit board, refer back to Chapter 31, "Repairing a VCR that has been dropped."

Reassembling a Remote Control

In older units:

1. Place the top front cover upside down and put all the plastic buttons back into their proper holes. Be sure that the buttons are inserted straight.

2. Place the rubber pad over the buttons with the conductors facing up.

3. Be sure that the proper end of the pad is facing toward the front. You can do this by placing the bottom side of the circuit board beside the pad to compare the conductors to their contacts, as shown in Figure 33.23.

4. Place the circuit board so that the contacts are facing the pad.

5. Starting at the front of the board, push the board down into its cover.

6. If latches are on each side of the top cover, push on the circuit board by each latch until the latch slides over the board, as shown in Figure 33.24. Do each latch in the same manner.

7. Place the back cover up against the front cover and squeeze them together until you hear a popping sound from both sides, indicating that the latches inside have locked into place. Replace the mounting screws, batteries, and battery door cover.

FIGURE 33.23

Comparing conductors to the contacts.

FIGURE 33.24

Pushing on the circuit board to latch each latch.

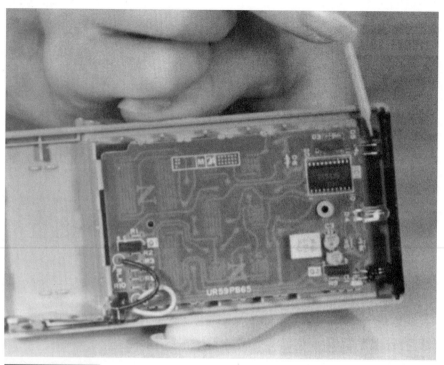

FIGURE 33.25

The base of an indicator lamp.

In some models, the remote has a red indicator lamp or LED on top of the unit. This lamp is mounted onto a circuit board. When remounting the circuit board, be sure that the lamp protrudes through its hole on the front cover, as shown in Figure 33.25. The lamp usually has long leads on it. These leads can get bent easily, causing the lamp to be misaligned. In newer models, the indicator lamps are incorporated onto the circuit board and there is nothing to line up.

In models with a red window:

1. Push it down into its proper position.

2. When remounting the back cover to the front cover, start by first sliding in the front end of the cover, as shown in Figure 33.26.

3. Drop the back cover down over the front cover and squeeze them together. You should hear a popping sound coming from each side. This means the latches inside are latched.

4. Replace any mounting screws to the back cover.

5. Replace the batteries and the battery cover.

FIGURE 33.26

Sliding the front end of the cover in first.

In models that have the remote with the buttons incorporated into the pad, place the button side of the pad back into the front cover. Look on the opposite side of the front cover to make sure that the buttons are inserted through their proper holes. Now, place the contact side of the circuit board over the pad while placing the LED through its proper hole in the front cover of the unit. If the unit has an indicator lamp, make sure it's properly inserted into the hole as you're fitting the circuit board back into its cover. Align the battery terminals with the grooves and push straight down (see Figure 33.12). Place the back cover up against the front cover and squeeze them together until you hear a popping sound from both sides.

In other models, place the battery springs through the bottom cover and leave the bottom cover overlapping the top cover by ¼ inch, as shown in Figure 33.27. Slide the bottom cover forward toward the battery compartment until all latches latch. Replace the mounting screws, batteries, and battery door cover.

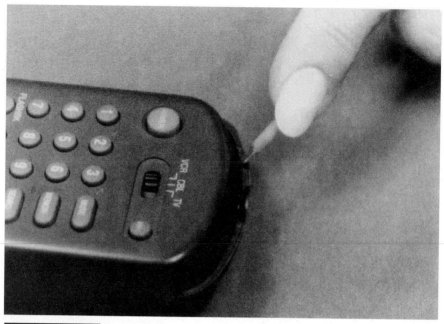

FIGURE 33.27

Bottom cover overlapping the top cover.

In some newer models, the circuit board is connected to the bottom cover and the pad is connected to the top cover (refer to Figure 33.16). Place the back cover up against the front cover and squeeze them together until you hear a popping sound from both sides. Replace the mounting screws, batteries, and battery door cover.

In the last type of remote, line up the circuit board with the pad and replace the circuit board mounting screws. Place both covers back to back, squeezing them together until you hear a popping sound coming from both sides. Then, replace any mounting screws, batteries and battery door cover.

Review

1. If the remote doesn't work, replace the batteries. If that doesn't work, then the unit must be disassembled for repair.

2. Look for any screws on the back cover, under the rubber foot pads, or inside the battery compartment.

3. When removing the back cover, use a small screwdriver to release the latches and/or use a twisting motion to pop the cover off.

4. Remove the circuit board by taking out the screws, by releasing the latches, or by prying the battery terminals up.

5. Check the circuit board for small cracks and repair them. Refer to Chapter 31.

6. Check the buttons for sticking. Clean the buttons and the inside of the holes in the front cover.

7. If you find a function button that doesn't work:

    A. Check the conductors on the rubber pads for dirt or beverages, then clean them.

    B. Check the contacts on the circuit board for dirt or beverages, then clean them.

8. To reassemble the remote, be sure that the rubber pads with the conductors are properly aligned to the contacts on the circuit board.

9. Be sure that all buttons protruding through the front cover are aligned in the proper location and aren't inserted crooked.

10. Be sure to put the battery springs back through the proper holes on the back cover.

11. Place both covers back to back and squeeze. Listen for a popping sound from each side of the remote.

Index

Illustrations are in **boldface**.

About the Author

Richard Wilkins created the concept for and wrote the bestselling first edition of *Home VCR Repair Illustrated*. He has been an electronics service technician for approximately 30 years, during which time he has owned and operated three successful businesses. He is now retired and traveling the United States with his wife Vicki.

Vicki McNairy-Wilkins has owned and operated a medical transcription business in Oregon and assisted with the writing and photography of this book. She is now retired and traveling the United States with her husband, Richard.